NJU SA 2019-2020

南京大学建筑与城市规划学院 建筑系教学年鉴
THE YEAR BOOK OF ARCHITECTURE DEPARTMENT TEACHING PROGRAM
SCHOOL OF ARCHITECTURE AND URBAN PL ANNING NANJING UNIVERSITY
胡友培 编　EDITOR : HU YOUPEI
东南大学出版社 · 南京　SOUTHEAST UNIVERSITY PRESS, NANJING

建筑设计及其理论
Architectural Design and Theory

张 雷 教 授	Professor ZHANG Lei
冯金龙 教 授	Professor FENG Jinlong
吉国华 教 授	Professor JI Guohua
周 凌 教 授	Professor ZHOU Ling
傅 筱 教 授	Professor FU Xiao
王 铠 副研究员	Associate Researcher WANG Kai
钟华颖 副研究员	Associate Researcher ZHONG Huaying
黄华青 副研究员	Associate Researcher HUANG Huaqing
梁宇舒 助理研究员	Assistant Researcher LIANG Yushu

城市设计及其理论
Urban Design and Theory

丁沃沃 教 授	Professor DING Wowo
鲁安东 教 授	Professor LU Andong
华晓宁 副教授	Associate Professor HUA Xiaoning
胡友培 副教授	Associate Professor HU Youpei
窦平平 副教授	Associate Professor DOU Pingping
刘 铨 副教授	Associate Professor LIU Quan
尹 航 讲 师	Lecturer YIN Hang
唐 莲 副研究员	Associate Researcher TANG Lian
尤 伟 副研究员	Associate Researcher YOU Wei

建筑历史与理论及历史建筑保护
Architectural History and Theory, Protection of Historic Building

赵 辰 教 授	Professor ZHAO Chen
王骏阳 教 授	Professor WANG Junyang
胡 恒 教 授	Professor HU Heng
冷 天 副教授	Associate Professor LENG Tian
史文娟 副研究员	Associate Researcher SHI Wenjuan
王丹丹 助理研究员	Assistant Researcher WANG Dandan

建筑技术科学
Building Technology Science

吴 蔚 副教授	Associate Professor WU Wei
郜 志 副教授	Associate Professor GAO Zhi
童滋雨 副教授	Associate Professor TONG Ziyu
梁卫辉 副教授	Associate Professor LIANG Weihui
施珊珊 副研究员	Associate Researcher SHI Shanshan
孟宪川 助理研究员	Assistant Researcher MENG Xianchuan
李清朋 博士后	Postdoctor LI Qingpeng

南京大学建筑与城市规划学院建筑系
Department of Architecture
School of Architecture and Urban Planning
Nanjing University
arch@nju.edu.cn   http://arch.nju.edu.cn

教学纲要
EDUCATIONAL PROGRAM

研究生培养（硕士学位）Graduate Program (Master Degree)			研究生培养（博士学位）Ph. D. Program
一年级 1st Year	二年级 2nd Year	三年级 3rd Year	

学术研究训练 Academic Research Training			

	学术研究 Academic Research		

建筑设计研究 Research of Architectural Design	毕业设计 Thesis Project	学位论文 Dissertation	学位论文 Dissertation

专业核心理论 Core Theory of Architecture	专业扩展理论 Architectural Theory Extended	专业提升理论 Architectural Theory Upgraded	跨学科理论 Interdisciplinary Theory

建筑构造实验室 Tectonic Lab
建筑物理实验室 Building Physics Lab
数字建筑实验室 CAAD Lab

生产实习 Practice of Profession   生产实习 Practice of Profession

课程安排
CURRICULUM OUTLINE

	本科一年级 Undergraduate Program 1st Year	本科二年级 Undergraduate Program 2nd Year	本科三年级 Undergraduate Program 3rd Year
建筑设计 Architectural Design	设计基础 Basic Design	建筑设计基础 Basic Design of Architecture 建筑设计（一） Architectural Design 1 建筑设计（二） Architectural Design 2	建筑设计（三） Architectural Design 3 建筑设计（四） Architectural Design 4 建筑设计（五） Architectural Design 5 建筑设计（六） Architectural Design 6
建筑理论 Architectural Theory		建筑导论 Introduction to Architecture 中国传统建筑文化 Traditional Chinese Architecture	建筑设计基础原理 Basic Theory of Architectural Design 居住建筑设计与居住区规划原理 Theory of Housing Design and Residential Planning 城市规划原理 Theory of Urban Planning
城市理论 Urban Theory			
历史理论 History Theory		中国建筑史（古代） History of Chinese Architecture (Ancient) 外国建筑史（古代） History of World Architecture (Ancient)	外国建筑史（当代） History of World Architecture (Modern) 中国建筑史（近现代） History of Chinese Architecture (Modern)
建筑技术 Architectural Technology	理论、材料与结构力学 Theoretical, Material & Structural Statics Visual BASIC程序设计 Visual BASIC Programming	CAAD理论与实践 Theory and Practice of CAAD	建筑技术（一）：结构、构造与施工 Architectural Technology 1: Structure, Detail and Construction 建筑技术（二）：建筑物理 Architectural Technology 2: Building Physics 建筑技术（三）：建筑设备 Architectural Technology 3: Building Equipment
实践课程 Practical Courses		古建筑测绘 Ancient Building Survey and Drawing	工地实习 Practice of Construction Plant
通识类课程 General Courses	数学 Mathematics 语文 Chinese 思想政治 Ideology and Politics 科学与艺术 Science and Art	社会学概论 Introduction of Sociology	
选修课程 Elective Courses		城市道路与交通规划 Planning of Urban Road and Traffic 环境科学概论 Introduction of Environmental Science 人文科学研究方法 Research Method of the Social Science 管理信息系统 Management Operating System 城市社会学 Urban Sociology	人文地理学 Human Geography 中国城市发展建设史 History of Chinese Urban Development 欧洲近现代文明史 Modern History of European Civilization 中国哲学史 History of Chinese Philosophy 宏观经济学 Macro Economics

本科四年级 Undergraduate Program 4th Year	研究生一年级 Graduate Program 1st Year	研究生二、三年级 Graduate Program 2nd & 3rd Years
建筑设计（七） Architectural Design 7 建筑设计（八） Architectural Design 8 本科毕业设计 Undergraduate Graduation Project	建筑设计研究（一） Design Studio 1 建筑设计研究（二） Design Studio 2 研究生设计工作坊 Graduate Design Workshop	建筑设计研究（三） Design Studio 3 专业硕士毕业设计 Thesis Project
	现代建筑设计基础理论 Preliminaries in Modern Architectural Design 研究方法与写作规范 Research Method and Thesis Writing	
城市设计理论 Theory of Urban Design 景观规划设计及其理论 Landscape Planning Design and Theory	城市形态研究 Study on Urban Morphology 城市形态与设计方法论 Urban Form and Design Methodology	
	建筑理论研究 Study of Architectural Theory	中国建构（木构）文化研究 Studies in Chinese Wooden Tectonic Culture
	材料与建造 Materials and Construction 计算机辅助建筑设计技术 Technology of CAAD	建筑环境学与设计 Architectural Environmental Science and Design 建筑技术中的人文主义 Technology of Humanism in Architecture
生产实习（一） Practice of Profession 1	生产实习（二） Practice of Profession 2	建筑设计与实践 Architectural Design and Practice
地理信息系统概论 Introduction of GIS 欧洲哲学史 History of European Philosophy 微观经济学 Micro Economics 建筑节能与绿色建筑设计 Building Energy Efficiency and Green Building Design	景观都市主义理论与方法 Theory and Methodology of Landscape Urbanism 建筑史研究 Studies in Architectural History GIS基础与运用 Concept and Application of GIS 传热学与计算流体力学基础 Foundamentals of Heat Transfer and Computational Fluid Dynamics 建筑环境学 Architectural Environmental Science 算法设计 Algorithmic Design 建筑体系整合 Building System Integration	建设工程项目管理 Management of Construction Project

2
设计基础
BASIC DESIGN

12
建筑设计（一）：独立居住空间设计
ARCHITECTURAL DESIGN 1: INDEPENDENT LIVING SPACE DESIGN

16
建筑设计（二）：文怀恩旧居加建设计
ARCHITECTURAL DESIGN 2: DESIGN FOR EXTENSION OF WEN HUAIEN'S FORMER RESIDENCE

20
中国传统建筑文化
TRADITIONAL CHINESE ARCHITECTURE

26
建筑设计（三）：幼儿园设计
ARCHITECTURAL DESIGN 3: THE KINDERGARTEN DESIGN

30
建筑设计（四）：书画家纪念馆
ARCHITECTURAL DESIGN 4: MUSEUM FOR ARTIST

34
建筑设计（四）：茶书吧设计
ARCHITECTURAL DESIGN 4: THE TEAROOM DESIGN

38
建筑设计（五）：大学生健身中心改扩建设计
ARCHITECTURAL DESIGN 5: RENOVATION AND EXTENSION DISIGN OF COLLEGE STUDENT FITNESS CENTER

48
建筑设计（六）：社区文化艺术中心设计
ARCHITECTURAL DESIGN 6: COMMUNITY CULTURE AND ENTERTAINMENT CENTER DESIGN

56
建筑设计（七）：高层办公楼设计
ARCHITECTURAL DESIGN 7: DESIGN OF HIGH-RISE OFFICE BUILDING

62
建筑设计（八）：城市设计
ARCHITECTURAL DESIGN 8: URBAN DESIGN

66
本科毕业设计：数字化设计与建造
UNDERGRADUATE GRADUATION PROJECT : DIGITAL DESIGN AND CONSTRUCTION

目 录

72
本科毕业设计：空间作为能动者：海上茶路的技术传播与东南亚近代聚落景观的塑造
UNDERGRADUATE GRADUATION PROJECT : SPACE AS AGENT: TRANSMISSION OF TEA TECHNOLOGY ALONG THE MARITIME TEA ROAD AND THE SHAPING OF MODERN SETTLEMENT LANDSCAPE IN EAST AND SOUTH ASIA

74
建筑设计研究（一）：基本设计
DESIGN STUDIO 1: DESIGN BASICS

80
建筑设计研究（二）：概念设计
DESIGN STUDIO 2: CONCEPTUAL DESIGN

86
建筑设计研究（三）：城市设计
DESIGN STUDIO 3: URBAN DESIGN

102
研究生国际教学交流计划
THE INTERNATIONAL POSTGRADUATE TEACHING PROGROM

110
研究生设计工作坊
GRADUATE DESIGN WORKSHOP

1-121 课程概览 COURSE OVERVIEW

123—134 建筑设计课程 ARCHITECTURAL DESIGN COURSES

135—137 建筑理论课程 ARCHITECTURAL THEORY COURSES

139—141 城市理论课程 URBAN THEORY COURSES

143—145 历史理论课程 HISTORY THEORY COURSES

147—150 建筑技术课程 ARCHITECTURAL TECHNOLOGY COURSES

151—157 其他 MISCELLANEA

课程概览
COURSE OVERVIEW

设 计 基 础
BASIC DESIGN

鲁安东 唐莲 尹航 梁宇舒

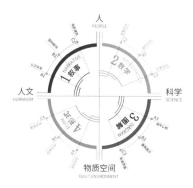

1.教案设计的背景
当代大学教育趋向于将博学与精专相统一的通识教育，建筑学的职业化教学体系需要同时满足通识教育的现代多元化育人要求。针对这一目标，本教案提出了新的设计基础教学体系，为创意工科大类提供核心的思维训练和能力培养。本课程为工科大类通识课，选课人数为80~100人，其中约三分之一后续选择建筑学方向。

2.培养多元融合的"设计思维"
本教案将创意工科中的"设计基础"理解为在人与物质空间之间，综合人文与科学多元途径的创造性实践，因此教学体系围绕着四条主题线索：

主题1　叙事：关注物质空间中个体人的感受和意义的视角。
主题2　数学：关注物质空间中人的共性和规律的视角。
主题3　图解：关注物质空间对人的容纳与支持的视角。
主题4　形式：关注物质空间本身及其运行状态的视角。

这四条主题线索各有侧重，帮助学生全面理解"设计"的本质，并为学生建立"设计思维"提供一个多元融合的整体视野和框架。

3.能力导向的培养路径
通识教育注重能力培养而不是技能培养。本教案将创意工科设计基础的核心能力分解为三个部分并引导学生逐步加以学习：

阶段1　感受认知能力：对个体人或者物质空间进行多角度、多形式的观察、记录、描述，关注对问题建立整体和理性的认识。
阶段2　分析转化能力：在感受认知的基础上，在人与物质空间、人文与科学的整体视野和框架下，关注对具体问题的分析与转化。
阶段3　创造设计能力：在认知与转化的前提下，创造性地提出、分析和处理问题，关注从思维到行动的实施与完成。

4.基于自主学习的模块化教学
本教案基于四条主题线索和三个能力培养阶段，设计了12个时长五周的教学模块。学生可以根据自己的兴趣和需求自由选修不同模块，量身塑造自己的设计思维和设计能力。通过模块化教学，本教案发挥了通识教育下自主学习的优势，开展理性、全面的思维训练，突出系统、多元的能力培养。通过将设计基础作为创意工科的"元"学科，既为学生进一步的专业学习打下扎实基础，也培养了学生未来跨学科创新的必要素质。

1.Design background of the teaching plan
Contemporary university education tends to adopt liberal education integrating extensive and specialized knowledge. The professional teaching system of architecture should also meet the requirements for modern diversified education of liberal education. In response to this objective, this teaching plan proposed a new basic design teaching system, which could provide core thinking training and competence training for creative engineering. This course, as a general course in engineering, may admit 80~100 students, about one-third of whom would select architecture in the future.

2.Cultivation of multivariate "design thinking"
In this teaching plan, the "basic design" in creative engineering is interpreted as creative practice integrating humanities and science between human and physical space; therefore, this teaching system centers on four thematic clues:

Theme 1 Narration: Pay attention to individual feeling and significance in physical space.
Theme 2 Mathematics: Pay attention to people's commonality and laws in physical space.
Theme 3 Illustration: Pay attention to the physical space itself and its operating state.
Theme 4 Form: Pay attention to accommodation and support of people in physical space.

These four thematic clues focus on basic design, with the aim of helping the students to fully understand the essence of "design", and providing a multivariate overall vision and framework for the establishment of "design thinking" among the students.

3.Competence-oriented training path
Liberal education focuses on competence training rather than skill cultivation. In this teaching plan, the core competence of basic design of creative engineering is divided into three parts, and the students are guided to engage in gradual learning:

Stage 1 Perception cognition competence: Observe, record, and describe individual people or physical space in multiple angles and forms, and focus on establishing a holistic and rational understanding of problems.
Stage 2 Analysis and transformation competence: On the basis of perception cognition, and under the overall vision and framework of human and physical space, as well as humanity and science, focus on the analysis and transformation of specific issues.
Stage 3 Creative design competence: Under the premise of cognition and transformation, creatively propose, analyze and deal with problems, and focus on implementing and completing the process from thinking to action.

4.Modular teaching based on autonomous learning
In this teaching plan, 12 five-week teaching modules are designed based on the four thematic clues and three competence training stages. The students can freely select different modules according to their own interests and needs, to tailor their design thinking and design competence. Through modular teaching, this teaching plan can make use of the advantages of autonomous learning under liberal education, thus carrying out rational and comprehensive thinking training, and highlighting systematic and multivariate competence training. Basic design, the "meta" subject of creative engineering, can lay a solid foundation for professional learning by the students, and cultivate the necessary competences for interdisciplinary innovation in the future.

感受认知	分析转化	创造设计
5周（个人作业） 每组20人左右	5周（个人作业） 每组20人左右	5周（个人作业） 每组20人左右
A1 鲁安东 空间中的身体 	B1 鲁安东 园林剧场 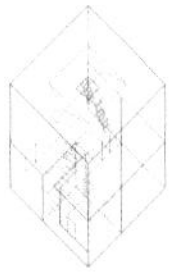	C1 鲁安东 日常影像的空间 
A2 唐莲 度量空间 	B2 唐莲 游历空间 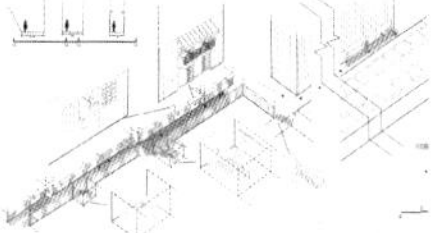	C2 唐莲 包裹空间 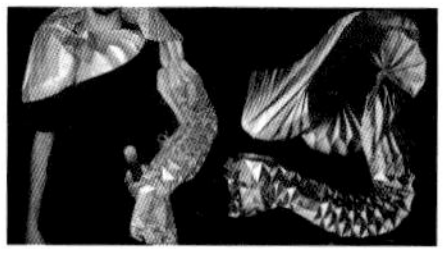
A3 梁宇舒 居住文化认知 	B3 梁宇舒 住屋原型分析 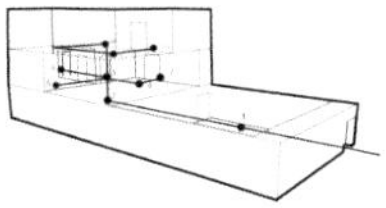	C3 梁宇舒 社区空间重构 
A4 尹航 街道界面认知 	B4 尹航 建筑空间认知 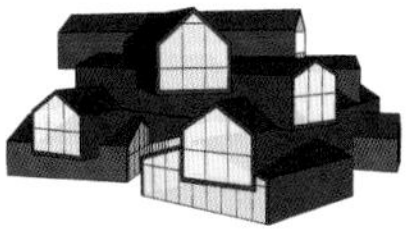	C4 尹航 城市认知图示 

设计基础 BASIC DESIGN

A1：空间中的身体
A1: BODY IN SPACE

鲁安东

教学进程

第一周
模块简介
讲课："空间中的身体 I"
课后作业：
身体空间1
分析一幅传统绘画（东西方均可）中的空间构成关系。

第二周
讲课："空间中的身体 II"
课后作业：
身体空间2
从网络搜索2张包含着不同身体空间关系的图片。使用Photoshop分析画面中身体和空间的关系。

第三周
课后作业：
身体动态
参考埃德沃德·迈布里奇的动作分解法，选取一个电影镜头或自拍一个镜头，分析一个动作的动态过程所需的空间。

第四周
课后作业：
身体情境
选取一个室内电影镜头，分析：
1）影像画面内的身体空间关系，制作一张合成分析图。
2）绘制房间平面图，分析摄像机和画面的关系。

Teaching process

Week 1
Module introduction
Lecture on Body in Space I
Homework:
Embodied Space 1
Analyze the spatial composition relationship of a traditional painting (eastern or western painting).

Week 2
Lecture on Body in Space II
Homework:
Embodied Space 2
Search two pictures containing different relationships of Embodied Space from the Internet, and analyze the relationships by Photoshop.

Week 3
Homework:
Body Dynamics
Referring to the action decomposition method of Eadweard Muybridge, select an image from a film or shoot an image, to analyze the space required for the dynamic process of an action.

Week 4
Homework:
Embodied Situation
Select an indoor image from a film:
1) Analyze the relationship of Embodied Space in the image, and prepare a synthetic analysis diagram.
2) Draw a plan of the room and analyze the relationship between the camera and the image.

设计基础 BASIC DESIGN

C1：日常影像的空间
C1:SPACE FOR DAILY IMAGE
鲁安东

教学进程
第一周
模块简介
讲课："日常性的影像博物馆"
课后作业：
自选三个电影片段，寻找电影中的一个日常建筑元素：楼梯、走廊、窗、门、阳台、露台、庭院，并配文描述建筑元素的空间特质。
第二周
上周作业讲评
课后任务解析
课后作业：
对建筑元素作出定义，运用建筑元素的重复，塑造一个无尽的空间。
第三周
上周作业讲评
课后任务解析
课后作业：
基于无尽的空间，叠加一种新的功能，运用建筑元素，将这个空间转变为建筑。
第四周
上周作业讲评
课后任务解析
课后作业：优化建筑形式，完善细节。

Teaching process
Week 1
Module introduction
Lecture on Daily Image Museum
Homework:
Select three movie clips, and look for daily architectural elements, such as stairs, corridors, windows, doors, balconies, terraces, and courtyards, and describe their spatial characteristics.
Week 2
Comment on homework of the last week
After-class task analysis
Homework:
Define the architectural elements, and create an endless space with repeated architectural elements.
Week 3
Comment on homework of the last week
After-class task analysis
Homework:
Based on the endless space, superimpose a new function and transform the space into a building with the architectural elements.
Week 4
Comment on homework of the last week
After-class task analysis
Homework: Optimize the architectural form and improve the details.

设计基础 BASIC DESIGN
B2：游历空间
B2:TOUR OF SPACE

唐莲

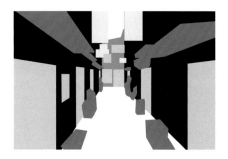

1.教学目标
培养城市空间感知能力，理解建筑与城市空间的关系，训练制图能力。

2.教学内容
"游历空间"的教学历时5周（含评图1周），通过分析与人的感知最为接近的街景视频，认知与分析城市空间。教学内容包括绘图、拼贴等：选取不同类型的城市游历街景视频（给定或自选），提取与绘制空间要素，再现游历空间意象。教学过程包括三个阶段的练习：街道空间街景切片提取与分析（1周），空间要素提取与绘制（2周），空间拼贴与表现（1~2周）。

3.教学进程
课前准备：硫酸纸、绘图笔等。软件：Photoshop，或Illustrator，sketchUp等
第一周
1）准备与介绍。
2）讲课："城市空间感知"。
3）课上练习（选择街景照片，对空间进行分析）。
第二周
1）课后作业讲评（电子版）。
2）讲课："城市空间意象"。
3）课上练习（选择街景照片，对空间要素进行分析）。
第三周
1）课后作业讲评：扫描合成图（电子版）。
2）讲课："城市空间图示方法"。
第四周
1）课后作业讲评：扫描合成图（电子版）。
2）讲课："城市空间感知的延展研究"。

1. Teaching objective
Cultivate urban space perception competence, understand the relationship between buildings and urban space, and train the drawing competence.

2. Teaching contents
This course lasts for five weeks (including drawing review for 1 week), aiming to cognize and analyze urban space by analyzing the street view video being closest to human perception. Teaching contents include drawing and collage: Select different types of street view videos (given or self-selected), extract and draw the space elements, to show the space images. The teaching process consists of three stages of exercises: Street view slice extraction and analysis (1 week), space element extraction and drawing (2 weeks), space collage and expression (1~2 weeks).

3.Teaching process
Preparation before class: Vegetable parchment, and drawing pen. software: Photoshop, or Illustrator, sketchUp, etc..
Week 1
1) Warm up & introduction.
2) Lecture on Urban Space Perception.
3) Class practice (Select photos of street view, and analyze space elements).
Week 2
1) Comment on homework(electronic edition).
2) Lecture on Urban Space Image.
3) Class practice (Select photos of street view, and analyze space elements).
Week 3
1) Comment on homework: Scanning composite diagram (electronic edition).
2) Lecture on Urban Space Illustration Method.
Week 4
1) Comment on homework: Scanning composite diagram (electronic edition).
2) Lecture on Urban Space Perception Extension Research.

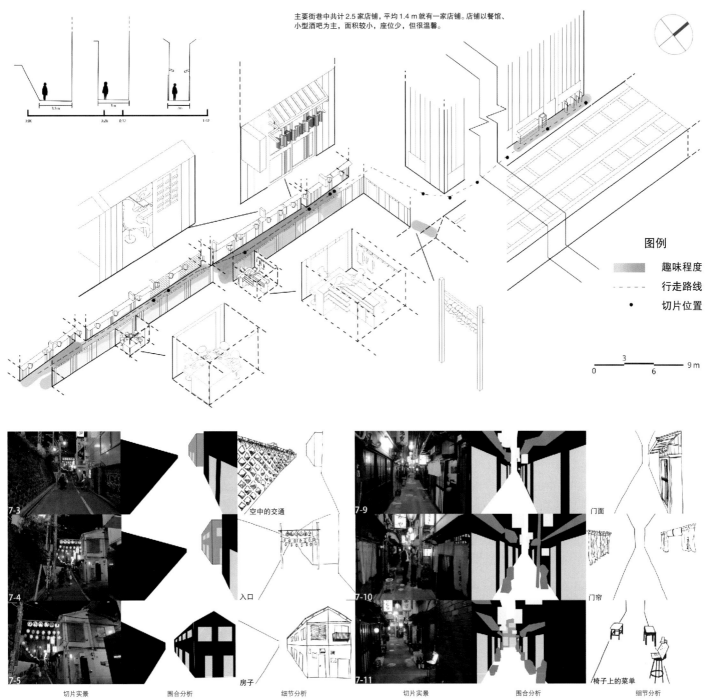

设计基础 BASIC DESIGN

C3：社区空间重构
C3: COMMUNITY SPACE RECONSTRUCTION

梁宇舒

1.教学内容

"社区空间重构"模块培养学生关注住屋与住屋之间的空间关系处理、路径处理、结构组织处理，鼓励再现原有聚落的文化信仰、地理特征、场所意向等多要素中的一个或多个，在给定框架内设计一座"新型社区"。

教学过程历时六周（含评图1周），包括四个练习：

A：一人一组，以教师提供的聚落样本为参考，选取一个聚落绘制（整体/局部）聚落形态总平面分析图；

B：一人一组，以前一阶段的单元设计为基础，制订建筑单元体的标准模块选形、块数计划，要有1~2个社区中心；

C：一人一组，制作场地模型（1:1000），根据地域要素、地理特征等保留一定的场地信息；

D：一人一组，针对典型聚落空间结构要素，抽取单一要素，在标准模块的基础上及空置的场地中设计一个小型未来社区。

2. 教学进程

第一周
1) 讲课："传统聚落探访"；2) 布置作业A。

第二周
1) 汇报讲评作业A（1h）；2) 讲课："当代社区的重构"；3) 布置作业B。

第三周
1) 汇报讲评作业B（1h）；2) 布置作业C。

第四周
1) 汇报讲评作业C（1h）；2) 布置作业D。

第五周
深化制作模型，汇总完善成果，准备评图。

1.Teaching contents

"Community space reconstruction" module is mainly used to cultivate the students to focus on the handling of spatial relationship between houses, the processing of path and the treatment of structural organization; and encourage them to reproduce one or several factors such as the cultural beliefs, geographic features, and site intentions of the original colonies, so as to design a "new community" within a given framework.

The teaching process lasts for six weeks (including drawing review for 1 week) and consists of four exercises:

A: One student in a group. Select a settlement to draw a general layout for (overall/partial) settlement based on the sample provided by the teacher.

B: One student in a group. Develop the standard module selection for building blocks and the planning of number of blocks based on the design of the previous stage, and there should be 1~2 community centers.

C: One student in a group. Develop a site model (1:1000), and keep certain information of the site according to territorial elements and geographic features.

D: One student in a group. Select a single element from the spatial structure elements of typical colonies, and design a small future community at a vacant site on the basis of the standard module.

2.Teaching process

Week 1
1) Lecture on Visit of Traditional Colonies; 2) Homework A.

Week 2
1) Comment on Homework A (1 h); 2) Lecture on Reconstruction of Contemporary Community; 3) Homework B.

Week 3
1) Comment on Homework B (1 h); 2) Homework C.

Week 4
1) Comment on Homework C (1 h); 2) Homework D.

Week 5
Deepen the model production, summarize and improve the results, and make preparations for the final review.

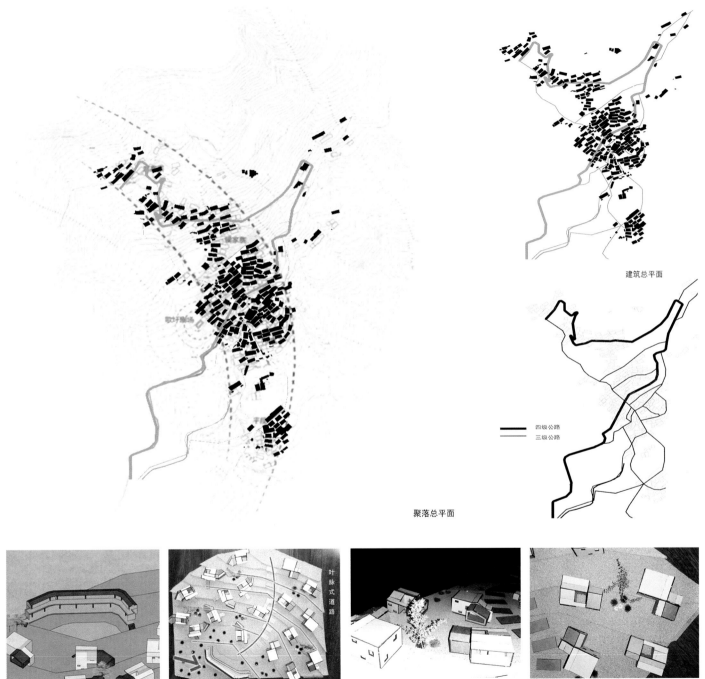

建筑总平面

聚落总平面

叶脉式道路

—— 四级公路
—— 三级公路

建筑设计（一）ARCHITECTURAL DESIGN 1

独立居住空间设计
INDEPENDENT LIVING SPACE DESIGN
刘铨 冷天 黄华青

1.教学目标
本次练习的主要任务是，综合运用在建筑设计基础课程中的知识点，初步体验一个小型独立居住空间的设计过程。训练的重点在于内部空间的整合性设计，同时希望学生在设计学习开始之初，能够主动去关注场地与界面、功能与空间、流线与出入口、尺度与感知等设计要素之间的紧密关系。

2.教学要点
1) 场地与界面：场地从外部限定了建筑空间的生成条件。本次设计场地是南京老城内的真实建筑地块，面积在80~100 m²左右，单面或相邻两面临街，周边为1~2层的传统民居。作为第一个设计训练，教案对场地环境条件做了简化限定，主要是要求学生从场地原有界面出发来考虑新建建筑的形体、布局及其最终的空间视觉感受。

2) 功能与空间：使用者的不同功能需求是建筑空间生成的主要动因，也是建筑设计要解决的基本问题。本次设计的建筑功能为小型家庭独立式住宅并附设有书房功能，家庭主要成员包括一对年轻夫妇和1位未成年儿童（7岁左右），新建建筑面积160~200 m²，建筑高度≤9 m（室外地面至女儿墙顶，不设地下空间）。设计者必须独立设定家庭主要成员各自的身份背景及兴趣爱好（如具有理、工、医、法等高等教育背景人士），并依此发展出内部不同功能性居住空间（包括但不仅限于起居室、餐厅、卧室、儿童房、厨房、厕所、浴室、储藏室等），并通过三维的空间设计探讨各居住空间之间的整合关系。

3) 流线与出入口：一方面，建筑内部各功能空间需要合理的水平、垂直交通来相互沟通与联系；另一方面，建筑的内部空间需要考虑与场地周边环境条件的合理衔接，如街道界面的连续性、出入口位置的选择与退让处理、周边建筑外墙界面（包括其上的外窗）对新建建筑的影响、建筑之间的间距与视线干扰、日照的合理使用等。新建建筑的内部楼梯必须符合现行国内住宅内部楼梯相关规范（踏面宽度22~28 cm，踏步高18~20 cm，两面靠墙时梯段宽度90 cm，一边临空时梯段宽度75 cm，连续踏步数量不得超过18级等）。

4) 尺度与感知：建筑内部的空间是供人来使用的，因此建筑中的各功能空间的尺度，都必须以人体作为基本的参照和考量，并结合人体的各种行为活动方式，来确定合理的建筑空间尺寸。在空间形式处理中注意通过图示表达理解空间构成要素与人的空间体验之间的关系，主要包括尺度感和围合感。

3.教学进程
本次设计课程共6周（2020.03.10—2020.04.21）
第一周：构思并撰写几个有代表性的生活场景（家庭人物构成、人物相对关系、类似戏剧中的"折"之剧本），利用基本的九宫格原形，构思概念性内部空间模块，及其关系。

第二周：用1：50手绘平、立、剖面图纸，在初步方案的基础上深化考虑功能与空间、流线与尺度。

第三周：利用1：50实体工作模型辅助设计，确定设计方案，推进剖、立面设计。
注：正式返校后，调研实际场地，分组制作场地模型（底座60 cm×60 cm×5 cm）。

第四周：深化1：50图纸，细化推敲各设计细节，并建模研究内部空间效果（集中挂图点评）。

第五周：制作1：20剖透视和各分析图，制作1：20大比例模型。

第六周：整理图纸、排版并完成课程答辩。

4.成果要求
A1灰度图纸2张，纸质表现模型1个（比例1：20），工作模型若干。图纸内容应包括：
1) 总平面图（1：200），各层平面图、纵横剖面图和主要立面图（1：50），内部空间组织剖透视图1张（1：20）。
2) 设计说明和主要技术经济指标（用地面积、建筑面积、容积率、建筑密度）。
3) 表达设计意图和设计过程的分析图（体块生成、功能分析、流线分析、结构体系等）。
4) 纸质模型照片与电脑效果图、照片拼贴等。

5.地块选择
可供选择的共有四块用地，其中①、②、③号地块均为单面临街，默认分配详见名单。

④号地块为双面临街，各小组内有对④号地块感兴趣的同学，可以向本组带队老师提出换地要求，否则保持默认地块。

1.Teaching objective
The main task of this exercise is to comprehensively use the knowledge points in basic courses of architectural design, and preliminarily experience the design process of a small independent living space. The training focuses on integrated design of internal space. At the beginning of design study, students are expected to actively pay attention to the close relationship between site and interface, function and space, circulation and access, dimension and perception.

2.Teaching points
1) Site and interface：The site may define the architectural space from the outside. The design site is a real plot located in the old town of Nanjing, with an area of about

 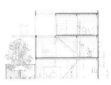 
 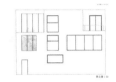 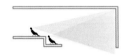

80~100 m^2, which faces the street on one side or two adjacent sides, with traditional houses of 1~2 storeys. As the first design training, the site environmental conditions are simplified in the teaching plan, which mainly asks the students to consider the shape, layout and final spatial visual experience of a new building from the original interface of the site.

2) Function and space: The main motivation for the forming of architectural space is the difference in function needs of the users, which is also a basic problem that should be resolved through architectural design. The function of this design is a detached house with a study of a small family. The family consists of a young couple and a child (7 years old or so). The house has a newly built area of 160~200 m^2, with the building height ≤9 m (the distance from outdoor ground to top of parapet, without underground space). The designers must independently set the identity background and hobbies of major members of the family (they should be personnel with higher education background, such as those in the fields of science, engineering, medicine and law), and then develop different functional living spaces (including but not only in the living room, dining room, bedroom, children's room, kitchen, toilet, bath room and storeroom); finally, they should discuss the integration relationship between various residential spaces based on three-dimensional space design.

3) Circulation and access: On the one hand, the function space in the building should be communicated and contacted by rational horizontal and vertical transportation; on the other hand, the internal space should be rationally connected to the surrounding environmental conditions, such as continuity of the street interface, the selection and concession of access, the influence of external wall interface (including the external windows) of the surrounding buildings on the new building, the distance between buildings and interference of vision, and the rational use of sunlight. The internal stairs of the new building must comply with the current national regulations on internal stairs (tread width: 22~28 cm, step height: 18~20 cm, stair width when the two sides are against the wall: 90 cm, stair width when one side is next to the air: 75 cm, number of consecutive steps: no more than 18)

4) Dimension and perception: The internal space of a building is for human use, so the dimension of each function space must take human body as basic reference and consideration, and various behaviors and activities must be combined to determine the rational dimension of architectural space. In space form processing, attention should be paid to understand the relationship between space components and spatial experience of people, mainly including the sense of dimension and enclosure.

3. Teaching process

This design course lasts for 6 weeks (March 10-April 21, 2020)

Week 1: Conceive and write several representative scenes of life (the composition of the family, the relative relationship of family members, and the script of a section in drama), and then determine the conceptual internal space modules and their relationships based on the basic Sudoku Model.

Week 2: Draw the details between function and space, circulation and dimension with a 1:50 hand-drawn plane, elevation and section based on the preliminary scheme.

Week 3: Assist the design with a 1:50 working model, to determine the design scheme, and promote the section and elevation design.

Note: After officially returning to school, the students will be arranged to investigate the site and develop the site model (pedestal: 60 cm×60 cm×5 cm) in groups.

Week 4: Deepen the 1:50 drawing, refine the design details, and set a model to study the effect of the internal space (concentrated comment on drawings).

Week 5: Draw the 1:20 perspective and analysis charts, and establish a 1:20 large-scale model.

Week 6: Organize the drawings, perform typesetting, and complete the course reply.

4. Achievement requirements

2 pieces of A1 grey-scale drawing, 1 paper presentation model (scale: 1:20), and several working models. Drawing contents should include:

1) General layout (1:200), plan of each storey, vertical and horizontal section and main façade drawing (1:50), 1 perspective view of internal space organization (1:20).

2) Design description and main technical and economic indicators (land area, building area, plot ratio, building density).

3) Analysis chart expressing design intent and design process (block generation, function analysis, circulation analysis, and structural system).

4) Paper model photos, computer renderings, and photo collages.

5. Lot selection

There are four lots to be selected, of which, Lot 1, 2 and 3 face the street with one side, and please see the list for default allocation.

Lot 4 faces the street with two sides, and the students in each group who are interested in Lot 4 may ask the teacher to make an exchange, otherwise, the default lot will be maintained.

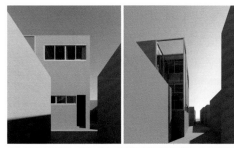

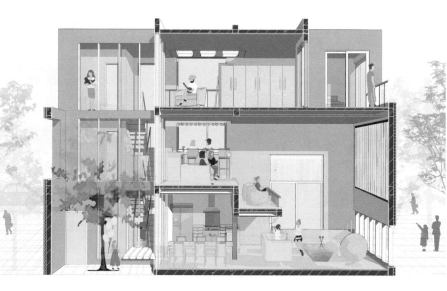

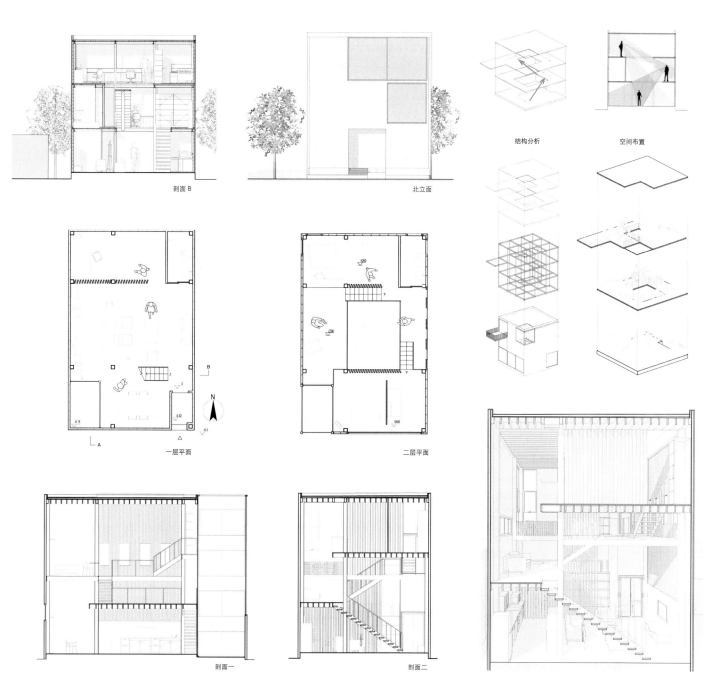

文怀恩旧居加建设计
DESIGN FOR EXTENSION OF WEN HUAIEN'S FORMER RESIDENCE
刘铨 冷天 黄华青

1.教学目标
上一个设计题目关注的重点是利用水平构件（楼板）组织和限定建筑内部空间，利用楼板的大小形状、开洞或错位关系形成所需的不同空间功能尺度与视觉联系。在本次设计中，我们需要进一步增加对场地环境的关注，注意新老建筑、内外空间、历史记忆与现实需求的关系，同时不仅利用水平构件，还要充分利用垂直构件（墙体、柱子）来组织和限定功能流线与营造视觉体验。

2.设计场地
设计训练的场地位于南京大学鼓楼校区新老轴线之间、教学楼东南侧文怀恩旧居所在的区域。文怀恩旧居总建筑面积约667 m²（含地下室67 m²，阁楼73 m²），由主楼与附楼两部分组成，主入口朝东，面对金陵大学主轴线和小纪念公园。设计范围位于旧居东侧，面积大约1337.22 m²，包括了目前的一条南北向人行通路和公园的一部分，新建建筑红线则在涉及范围的北侧，面积约451.20 m²（详见地形与红线图）。

3.设计要求
1) 功能要求：根据文怀恩与金陵大学的历史，设计一个展示纪念馆。老建筑由于缺少大空间，因此需要加设一个较大的灵活空间。建成后老建筑用作固定展陈和办公，新建筑则作为临时展览、研讨交流、会议茶歇等可以灵活使用的空间。同时由于其位于校园核心地带，新增建筑内拟设一个小型咖啡厅，服务学校教职工及日常参观人群。新建建筑总面积不少于150 m²，建筑高度≤8 m（檐口高度，不包括女儿墙），新建建筑以一层为主，可局部夹层。在场地内还应考虑一处与展示主题相关的纪念性空间。新老建筑应作为一个整体考虑其参观流线，但新建建筑也应考虑其相对独立性，在老建筑闭馆时可独立使用。

2) 场地环境：现状建筑和场地内各项要素既为限制，又是形成新建筑体量的基本条件。本次设计场地南侧为小礼拜堂，北侧为教学楼，东侧面对小花园（原金陵大学主轴线上的入口花园）和金陵大学主轴线上的石碑、雕塑。结合与这些重要的环境要素的位置的视觉关系，考虑建筑物、纪念性空间的布局以及参观流线的组织。

3) 空间限定要素与视觉关系的组织：本次训练需要通过空间限定要素（水平与垂直构件）与身体感知关系的组织（路径、视线、活动与尺度、光影、质感），塑造出相应的室内外展陈与纪念性空间，连接文怀恩旧居历史记忆与现实需求，创造性地再现该场所的人文内涵。

4) 材料与建造：选择合适的材料、结构形式，呼应空间组织与场地环境需要。

4.教学进程
第一周：通过案例分析，了解复兴历史建筑的改建加建操作思路和具体策略。进行空间组织训练。

第二周：场地认知，结合已有图像资料、地形资料进行深度的场地认知。制作比例1∶50的场地模型。自拟文怀恩旧居的校园新功能，写出详细的展陈和使用设定，包括活动场所服务对象的类别和数目，场地设施的具体要求，思考建筑策略的灵活性。

第三周：提出处理文本设定相应的建筑策略，思考场地的新旧对话关系，整合空间的组织方式，形成初步方案与工作模型，比例1∶100。

第四周：深化初步方案，优化并发展前述的策略，用1∶50的手绘平、立、剖面图纸，在初步方案的基础上深化功能与空间、流线与尺度（集中挂图点评）。

第五周：利用工作模型辅助设计，进一步优化设计，使得建筑结构清晰、明确、可认知，并研究内部空间效果，确定最终的设计方案。

第六周：深化1∶50图纸，细化推敲各设计细节，制作比例1∶50的模型（模型点评）。

第七周：方案优化，思考并选择图面表达的效果，制作必要的分析和效果图。

第八周：整理图纸、排版，制作正式模型并完成课程答辩。

5.成果要求
1) 总平面图（1∶300），一层场地平面图（包含文怀恩宅）（1∶100）、各层平面图、纵横剖面图和主要立面图（1∶50），内部空间组织剖透视图1张（1∶20）。

2) 设计说明和主要技术经济指标（用地面积、建筑面积、容积率、建筑密度）。

3) 表达设计意图和设计过程的分析图（体块生成、功能分析、流线分析、结构体系等）。

4) 纸质模型照片与电脑效果图、照片拼贴等。

1.Teaching objective
The last design theme focused on organizing and limiting the internal space of a building with horizontal components (floor slabs), and forming the required different spatial functional dimensions and visual connections by the size, shape, opening or dislocation of the floor slabs. In this design, we should further increase the attention to site environment, and the relationship between new and old buildings, internal and external space, historical memory and actual needs; meanwhile, we should make use of the horizontal components, and also the vertical components (walls, columns) to organize and limit the functional flow and create visual experience.

2.Design site
The site for design training is located between the old and new axes of Gulou Campus, Nanjing University, in specific, the area at the southeast of the teaching building where the Wen Huaien's Former Residence is located. The residence, with the overall building area of 667 m² (including the basement of 67 m², and attic

of 73 m^2), consists of the main building and attached building. With east entrance as the main entrance, it faces the main axis of Jinling University and a small memorial park. The design area (area: about 1,337.22 m^2) is located at the east side of the residence, which contains a north-south pedestrian and a part of the park. The red line of the newly-built building is on the north of the involved area, with an area of about 451.20 m^2 (See details in the topographic map and red line map).

3.Design requirements

1) Functional requirements: Design an exhibition hall based on the history of Wen Huaien and Jinling University. Due to lack of large space, the old building should be attached with a larger flexible space. The completed old building will be used for exhibition and office, and the new building will be used for temporary exhibition, seminars and exchanges, conferences and tea breaks. At the same time, due to its location in the core area of the campus, it is planned to set a small coffee shop in the new building, serving the teaching and administrative staff and daily visitors. The total area of the new building should not be less than 150 m^2, with the building height ≤8 m (the height of cornice, excluding the parapet). The new building is mainly one storey, with an interlayer in some parts. A commemorative space related to the theme of exhibition should be arranged. The visiting flow should be determined through taking the new and old buildings as a whole, but the independence of the new building should be considered, so as to ensure that it can be used when the old building is closed.

2) Site environment: The existing building and various elements of the site are limitations and basic conditions for the formation of new building volume. In this design, there is a chapel in the south, a teaching building in the north, and a small garden in the east (the original entrance garden on the main axis of Jinling University), as well as the steles and sculptures along the main axis of Jinling University. Combining the visual relationship with the location of these important environment elements, consider the layout of buildings and monumental space, and the organization of the visiting flow.

3) Organization of the relationship between spatial limitation elements and vision: In this training, the students should create corresponding indoor and outdoor exhibition and commemorative space based on the organization of the relationship between spatial limitation elements (horizontal and vertical components) and body perception (path, vision, activity and scale, light and shadow, texture), so as to link the historical memory to actual needs, and creatively reproduce the humanistic connotation of the place.

4) Materials and construction: Select appropriate materials and structural form that meet the requirements of space organization and site environment.

4.Teaching process

Week 1: Understand the operating ideas and specific strategies for the reconstruction and extension of historical buildings through case analysis. Perform spatial organization training.

Week 2: Site cognition. Deepen site cognition in combination with the existing images and topographic data. Prepare a site model (scale: 1∶50). Self-design its new functions, and determine the detailed exhibition and application setting, including the type and number of serving targets, the specific requirements of site facilities, and consider the flexibility of building strategies.

Week 3: Propose the handling text and set the corresponding architectural strategies, consider the dialogue between the new and the old buildings, integrate the organization form of space, and form the preliminary scheme and working model with the scale of 1∶100.

Week 4: Deepen the preliminary scheme, optimize and develop the aforementioned strategies, and deepen function and space, streamline and dimension based on the preliminary scheme with the 1:50 hand-drawn plane, elevation and section (concentrated comment on drawings).

Week 5: Assist the design with the working model, further optimize the design, make the building structure clear and cognizable, study the effects of internal space, and determine the final design plan.

Week 6: Deepen the 1∶50 drawing, deliberate the design details, and develop a model with the scale of 1∶50 (model review).

Week 7: Scheme optimization. Consider and select the effect of the graphic representation, and prepare the necessary analysis chart and rendering.

Week 8: Settle the drawings, perform typesetting, prepare the formal model and complete the course reply.

5.Achievement requirements

1) General layout (1∶300), a plan of the first storey (including the residence) (1:100), plan of each storey, vertical and horizontal section and main elevation (1∶50), and 1 perspective view of the internal space organization (1∶20).

2) Design description and main technical and economic indicators (land area, building area, plot ratio, building density).

3) The analysis chart expressing design intent and design process (block generation, function analysis, Circulation analysis, and structural system).

4) Paper model photos and computer rendering, photo collage.

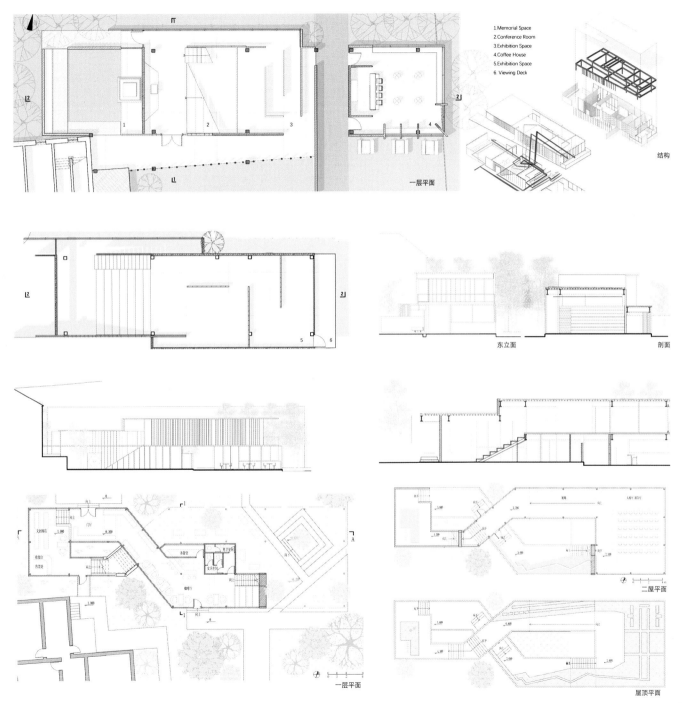

中国传统建筑文化
TRADITIONAL CHINESE ARCHITECTURE

本科二年级 UNDERGRADUATE PROGRAM 2ND YEAR

赵辰 史文娟

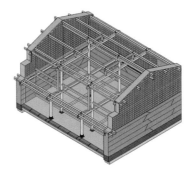

1 "中国房子"（基本单元）的营造设计
1 Construction and Design of "Chinese House" (Basic Unit)

1.教学目标
了解中国传统建造体系的基本单元的建造材料、建造程序，理解并掌握其建造的规律，乃至可适应不同地域的建筑形式。

2.教学方法
数字三维模型，模拟营造系统。

3.课程要求
第一阶段，根据提供的"中国房子"基本单元之平、立、剖图纸的数据信息，以及各个地域性建造体系的信息，建立"中国房子"的四种案例模型：慈城、剑川、闽东北、徽州。
第二阶段，将以建立的"中国房子"模型，根据体形改变的要求进行调整。

4.相关设计要素
1）尺度
面阔：3开间；10~12 m。
进深：步架数，屋面形体变化（屋檐、屋脊）；5~6步架；步架单位尺度根据地域规律。
高度：1~2层，底层3~4 m，二层3 m（檐口高度）。
2）构架
木构屋架：进深方向的柱、梁枋构成之各榀屋架；
　　　　　面阔方向的梁枋、檩（桁）条联系各榀屋架。
屋面：木构椽子；屋面板。
围合：墙体（夯土、砖石）。
各部位墙体：山墙、檐墙、院墙。
开启：门洞（门头）、窗洞。
3）屋面
硬山：山墙的各种变体（地域差异）和构造（排水）。
悬山：木构的出挑（檩条从山墙的悬出）。
瓦作：仰瓦、底瓦；勾头滴水。
屋脊（各种屋脊）：正脊、戗脊、垂脊。
5.教学进程
以"中国房子"的四种案例模型均分为四组：A，慈城组；B，剑川组；C，闽东北组；D，徽州组。由研究生助教指导建模，三周（2019年3月28日—4月18日）后答辩交作业。

1.Teaching objective
Understand the construction materials and procedures of the basic units of traditional Chinese construction system; understand and master the construction rules, and architectural forms adaptable to different regions.

2.Teaching method
Digital 3D Model, Simulation Building System.

3.Course requirements
At the first stage, establish four case models of "Chinese house": Cicheng, Jianchuan, Northeast Fujian, Huizhou, according to the data of the plane, elevation and sections of the basic unit, and the information of various regional construction systems.
At the second stage, make adjustments of the established "Chinese house" according to the change requirements.

4.The relevant design element
1) Dimension
Building width: 3 standard rooms; 10~12 m.
Depth: Number of purlins, changes in roof shape (eave, ridge); 5~6 purlins; the purlin unit dimension should be determined according to the geographic pattern.
Height: 1~2 storeys, 3~4 m of the first storey, 3 m of the second storey (cornice height).
2) Framework
Wooden roof truss: Roof truss formed by columns and beams in the direction of depth; roof truss formed by beams, purlins (stringers) in the direction of building width.
Roof: Wooden rafters; roof boarding.
Enclosure: Wall (rammed earth, masonry).
Walls of various parts: Gable, eave wall, courtyard wall.
Opening: Door opening (door head), window opening.
3) Roof
Flush gable roof: Various variants of gables (regional differences) and structures (drainage).
Overhanging gable roof: The outburst of wooden structure (purlin out of the gable).
Tilework: Top tiles, bottom tiles; triangle-shaped edge of the eave tile.
Ridge (various ridges): Main ridge, diagonal ridge and vertical ridge.
5. Teaching process
Divide the four case models of "Chinese house" into four groups: A. Cicheng Group; B. Jianchuan Group; C. Northeast Fujian Group; D. Huizhou Group. Modeling is guided by a graduate teaching assistant, and homework should be submitted after three weeks (March 28-April 18, 2019).

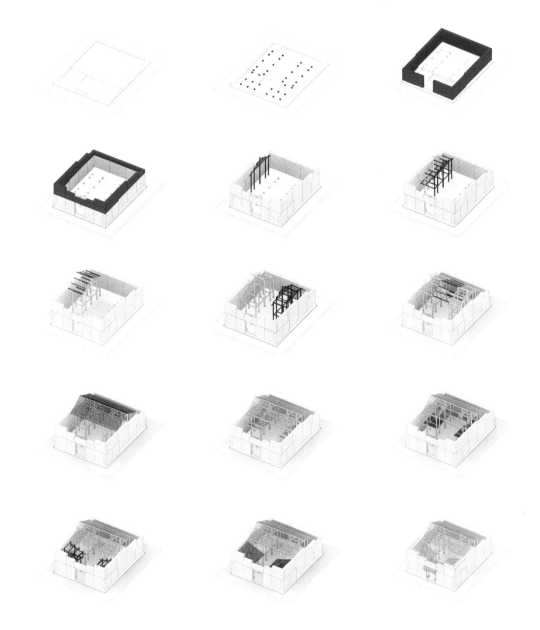

本科二年级 UNDERGRADUATE PROGRAM 2ND YEAR

中国传统建筑文化
TRADITIONAL CHINESE ARCHITECTURE

赵辰 史文娟

2 "中国院子"（空间组合体）的设计
2 Design of "Chinese Courtyard" (Space Combination)

1.教学目标
了解中国传统院落空间组合的基本建筑群体规划设计原则和方法，理解其传统的生活起居行为之内容及其建筑类型与形式的规律。

2.教学方法
在原有建构文化阶段完成的建筑基本单元的基础上，加以新设定的构成要素，组成空间组合的数字三维模型，并以图示表现。

3.教学内容
策划设定明清之际，江南传统城市社会某大户人家生活起居之行为模式，按"中国院子"的基本建筑空间组合的规律，选择小型、大型传统院落组合建筑群进行规划设计。

4.相关设计要素
1) 尺度
通面阔：路：中路、东西路；中路面阔为主，边路相应减之。
总进深：多进（两进以上）院落为单元，纵向发展；每一进单体的进深以主体建筑为主，特别是步架数和进深尺度；其余随之。
2) 院落
开敞空间的院子：因所处区位而定义，尺度、形式、材料。
开敞空间院落的变体：庭院；天井；偏院。
主体建筑：相应院落为主体。
3) 围合
门屋、倒座、厢房、院墙等成为院落空间的围合边界。与其他建筑单体共同构成院落形态。
4) 配房
厢房：与正房相配合纵向发展。
耳房：附属正房或厢房的偏屋。
灶间：工房；杂院。
5) 入口
主入口：礼仪性出入口，形势变化较多，尺度较大，双开门。
后门：主入口的背向，生活性出入口，结合灶间、工房、杂院。
边门（便门）：侧门，结合廊道，院墙。

1. Teaching objective
Understand the planning and design principles and methods of basic building groups of the combination of traditional Chinese courtyard space, understand the contents of traditional daily life, and the rules of architectural types and forms.
2. Teaching method
On the basis of basic building units completed at the original construction culture stage, attach new constituent elements to form a digital three-dimensional model of spatial combination, and express it with diagrams.
3. Teaching contents
Plan and set the behavior pattern of a large family in the traditional urban society of Jiangnan in the Ming and Qing Dynasties, and select small and large traditional courtyard combination buildings for planning and design based on the combination rules of the basic architectural space of "Chinese Courtyard".
4.The relevant design elements
1) Dimension
General width: Row: middle road, east-west road; mainly the width of middle road, the side roads are reduced accordingly.
Total depth: Multiple (more than two) courtyards form a unit, developing longitudinally; the depth of a single unit is formed by the main building, mainly the number of purlins and depth, followed by other parts.
2) Courtyard
Yard with open space: Dimension, form and materials should be defined by location.
Variations of air yard with open space: Courtyard; patio; side yard.
Main building: The corresponding courtyard.
3) Enclosure
Gate house, the house facing north, wing room, and courtyard wall form the boundary of the courtyard, and form the courtyard with other buildings.
4) Attached rooms
Wing room: Vertical development with the principal room.
Appentice: The side room attached to the principal room or wing room.
Kitchen: Workshop; side yard.
5) Entrance
Main entrance: Ceremonial entrance, in various forms and large dimension, with double door.
Back entrance: The back of the main entrance, living entrance, combined with kitchen, workshop, and side yard.
Wicket door: Side door, combined with corridors and courtyard walls.

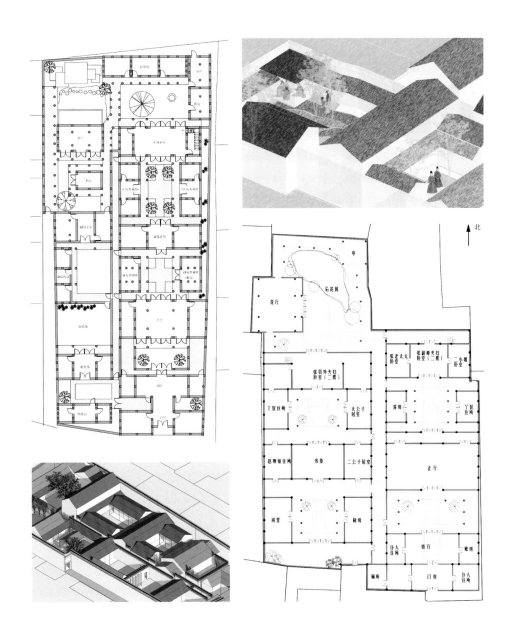

中国传统建筑文化
TRADITIONAL CHINESE ARCHITECTURE

赵辰 史文娟

3 "中国园子"（传统城市形态中的造园）设计
3 Design of "Chinese Garden" (Gardening in Traditional Urban Form)

1.教学目标
了解中国传统城市形态中的私家园林，以"中国院子"衍生而作为传统人居文化的一部分，进而理解从院落到街廓的传统城市形态规律。

2.教学方法
在以"中国院子"作业的基础上，加之家庭行为的休闲、文娱、养生内容的需求而空间衍生，并在新增的用地上以山水美学的意境构筑为庭园——"中国园子"。

3.教学内容
以传统院落空间组合为原则，以城市形态学为研究方法，结合中国城市的社会组织规律（社会结构、"地界"），将城市的可能的用地改造成私家园林，以满足亲近自然的人居行为。

4.相关设计要素
1）山水之山石、池沼。
2）建筑之堂榭、亭、廊、桥、墙等。
3）植栽之林、木、草。

5.基本设计程序
1）确定出入口位置（与城市、家宅之间的联系）。
2）明确园林主景（山景？水景？）以确立基地起伏态势与大致格局。
3）确立主建筑位置（既可为观景地点，亦可为被观之景）。
4）确立交通流线，配合廊、桥、山石小品、植栽等，以曲折有致。

6.最终成果表达
1）为自己所设计的园子和其中的景点赋名（点景）。
2）园子总平面。
3）结合住宅平面与文本中的人物与活动，给出游览动线的几种可能。
4）园景效果图创作（内容与表现形式自选）。
①以模型轴测为基础，绘制整体园景图。
②主要建筑中观赏到的主景或者游览过程中的景色（在模型或平面里标注观景点）。
③结合文本，园林中人物活动的场景图（在模型或平面里标注场景区域，鸟瞰、人视皆可）。

1. Teaching objective
Understand private gardens in the traditional Chinese urban morphology, and take it as the derivation of "Chinese Courtyard" and a part of traditional residential culture, thus understanding the rules of traditional urban morphology from courtyard to street profile.

2. Teaching method
Based on the "Chinese Courtyard", derive the space according to the leisure, entertainment, and health preservation demands, and create "Chinese Garden" with the artistic conception of landscape aesthetics in the newly-added land.

3.Teaching contents
Based on the principle of the combination of traditional courtyard space and the research method of urban morphology, and combined with social organization law of Chinese cities (social structure, "territory"), transform the possible land into private gardens, to realize the living close to nature.

4.The relevant design elements
1) Hill stone and ponds in terms of landscape.
2) Pavilions, corridors, bridges, and walls of buildings.
3) Forest, wood, grass of planting.

5.Basic design procedure
1) Determine the location of entrances and exits (connection with cities and homes).
2) Clarify the main landscape (mountain scenery? water scenery?), to establish the topography, and general pattern.
3) Determine the location of the main building (the viewing location or the location to be viewed).
4) Determine the traffic flow, and create twists and turns in combination with corridors, bridges, rocks and plants.

6.Final achievement expression
1) Assign names (point the scenes) to the designed garden and scenic spots.
2) General layout of the garden.
3) Possible tour lines in combination with house plan and the characters and their activities.
4) Creation of landscape rendering (optional content and form of expression).
① Based on the model axonometric diagram, draw the overall landscape map.
② The main scenery of the main building, or scenes viewed during the tour (mark them on the model or plane).
③ The scene of human activities in the garden combined with the text [mark the area (bird view/man view) on the model or the plane].

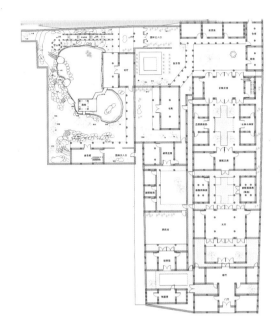
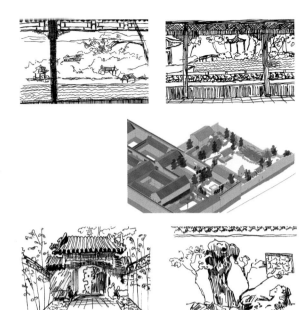

幼儿园设计
THE KINDERGARTEN DESIGN

童滋雨 华晓宁 窦平平

1. 教学目标

此课程训练解决建筑设计中的一类典型问题：标准空间单元的重复和组合。建筑一般都是多个空间的组合，其中一类比较特殊的建筑，其主体是通过一些相同或相似的标准空间单元重复而成，这种连续且有规律的重复，很容易表现出一种韵律和节奏感。对这类建筑的设计练习，可以帮助学生了解并熟悉空间组合中的重复、韵律、节奏、变化等操作手法。

2. 基地与任务

某幼儿园，用地面积约7200 m²。拟设托班、小班、中班、大班各3个，共计12个班，每班25人。使用面积约为2600 m²。高度不超过3层。

幼儿生活活动用房：活动室、寝室、卫生间、储藏室、音体室、图书室、科学室。

服务用房：晨检室、保健观察室、警卫室、教师值班室、储藏室、办公室、会议室、教具制作室、卫生间。

供应用房：厨房、休息更衣室、消毒开水间、洗衣间。

3. 室外场地设置要求

1) 每班应设专用室外活动场地，面积不宜小于60 m²，宜与班活动室毗连设置，各班活动场地之间宜采取分隔措施。

2) 应设全园共用活动场地，人均面积不应小于2 m²；要求布置集体活动场地17 m×17 m（可包括集体操场）和一组（4条）30 m长直跑道。

3) 地面应平整、防滑、无障碍、无尖锐突出物，并宜采用软质坪地。

4) 室外活动场地应有1/2以上的面积在标准建筑日照阴影线之外。

5) 场地内绿地率不应小于30%，宜设置集中绿化用地。

6) 供应区内宜设杂物院，并与其他部分相隔离。杂物院应有单独的对外出入口。

7) 幼儿生活用房应布置在当地最好朝向，冬至日底层满窗日照不应小于3 h。

8) 夏热冬冷地区的幼儿生活用房不宜朝西向；当不可避免时，应采取遮阳措施。

4. 成果要求

1) 空间与环境：总平面图（1：500）。

2) 空间基本表达：平、立、剖面图（1：200）。

3) 空间细化：幼儿活动室+寝室单元平面图（1：50）、墙体剖面图（1：50）。

4) 空间解析与表现：轴测图、分析图、剖透视和人眼透视。

5) 手工模型：1：200建筑模型。

5. 设计进度（共8周）

第1阶段：幼儿园设计资料收集归纳，设计幼儿活动室+寝室单元（平面图+剖面图）（1周）。

第2阶段：场地调研，单元组合设计（体块模型）（1周）。

第3阶段：深化设计（3周）。

第4阶段：单元放大设计，包括家具布置和构造设计（1周）。

第5阶段：最后成果制图、排版，准备答辩（2周）。

   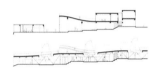

1. Teaching objective

The course training aims to solve a typical problem in architectural design: The repetition and combination of standard space units. A building is generally a combination of multiple space units. There is a type of special buildings, whose principal parts are composed by repetition of some identical or similar standard space units. Such a continuous and regular repetition can easily reflect a sense of rhythm. The exercises in design of such buildings can help the students to understand and familiarize the manipulation techniques for spatial combination such as repetition, rhythm and variation.

2. Base and task

Design a kindergarten, with the land area of about 7,200 m². It is planned to set up three nursery classes, three younger classes, three middle classes, and three top classes, 12 ones in total, with 25 children each. The usable area is about 2,600 m², and its height should not exceed 3 storeys.

Living and activity rooms: Activity room, bedroom, bathroom, storage room, music and sports room, library, and science room.

Service rooms: Morning inspection room, health observation room, guard room, teacher duty room, storage room, office, meeting room, teaching aid production room, and toilet.

Supply rooms: Kitchen, locker room, disinfection water heater room, and laundry room.

3. Outdoor venue setting requirements

1) Each class should set a dedicated outdoor activity venue, with the area of not less than 60 m², and it should be close to the activity room; the venues of different classes should be separated.
2) There should be a common activity venue, with the area per capita of no less than 2 m²; there should be a collective activity field of 17 m×17 m (including the playground) and a group of 30 m straight tracks (4 tracks).
3) The ground shall be smooth and non-slip without any barrier or sharp protrusions, and should be paved by soft material.
4) More than 1/2 of the outdoor activity venue should be outside the standard building sunshine shadow line.
5) The greening rate should not be less than 30%, and centralized greening land should be set.
6) The supply area should be set with a sundries yard, which should be isolated from other areas. The yard should have a separate entrance.
7) The children's living rooms should be arranged in the best orientation, and the full-window sunlight at the bottom floor should not be less than 3 hours in winter.
8) In regions with hot summer and cold winter, the children's living rooms should not face west; and when it is unavoidable, sun-shading measures should be taken.

4. Achievement requirements

1) Space and environment: General layout (1∶500).
2) Basic expression of space: Plane, elevation and section (1∶200).
3) Space refinement: Plane of children's activity room + bedroom unit (1∶50), wall section (1∶50).
4) Space analysis and performance: Axonometric drawing, analysis chart, sectional perspective and eye perspective.
5) Manual model: 1∶200 architectural model.

5. Design schedule (8 weeks in total)

Stage 1: Collect and sort design materials of the kindergarten, and design the children's activity room + bedroom unit (plane+section)(1 week).
Stage 2: Site investigation, unit combination design (block model)(1 week).
Stage 3: Deepen the design(3 weeks).
Stage 4: Start unit enlargement design, including furniture layout and structure design(1 week).
Stage 5: Perform final drawing and typesetting, and prepare the reply(2 weeks).

A-A 剖面

南立面

1-1 剖面

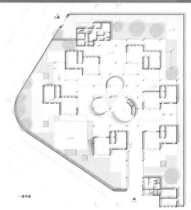

一层平面

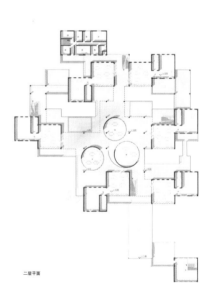

二层平面

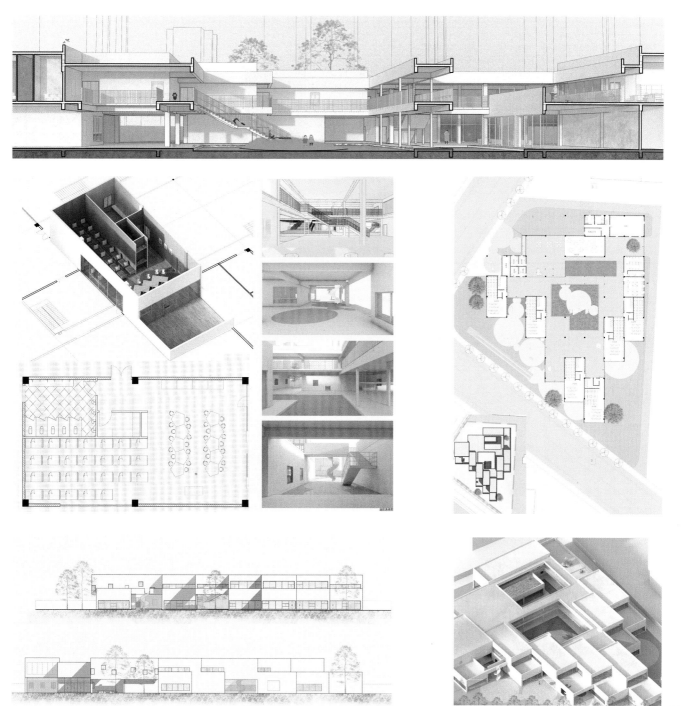

建筑设计（四） ARCHITECTURAL DESIGN 4
书画家纪念馆
MUSEUM FOR ARTIST

华晓宁 窦平平 孟宪川

1. 教学目标

课程主题是"空间"，学习建筑空间组织的技巧和方法，训练空间的操作与表达。

空间问题是建筑学的基本问题。课题基于复杂空间组织的训练和学习，从空间秩序入手，安排大空间与小空间，独立空间与重复空间，区分公共与私密空间、服务与被服务空间、开放与封闭空间。同时，充分考虑人在空间中的行为、空间感受，尝试以空间为手段表达特定的意义和氛围，最终形成一个完整的设计。

2. 基地与任务

作为十朝古都，南京历来是人文荟萃、名家辈出之地。故拟在长江路历史文化街区建造一座知名书画家的纪念馆，以促进社会文化事业的发展。学生自行选择一位历史上出生于南京，或是长期生活在南京，或是与南京有密切关联的著名书画家，分析其书画风格特质或个人品格，挖掘出纪念性主题，引领建筑空间的组织与设计。

基地位于江苏美术馆老馆北侧，碑亭巷与石婆婆庵交叉口西南角，用地面积约4100 m^2。

新建筑应具备以下功能：
1）展示书画家生平、事迹、作品、影响等；
2）收藏与该书画家相关的资料、档案、文物、作品等；
3）研究与该书画家相关的历史、理论、创作等；
4）艺术普及和社会教育功能，如举办艺术讲座、研讨会、沙龙等。

3. 空间计划

1）规划限定

建筑总高度不超过24 m，容积率不超过1.2。

2）空间组织原则

空间组织要有明确特征，有明确意图，概念要清楚，并且满足功能合理、环境协调、流线便捷的要求。注意不同类型、不同形态空间的构成、空间的串联组织和空间氛围的塑造。

3）空间类型

总建筑面积：4500 m^2（允许误差5%）。

（以下各部分面积配比为参考，每人可以根据研究自行策划进行适当调整）

①展览陈列空间：约2000 m^2。

展厅（或展廊）：1500 m^2（可设数个展区，含临时展厅）。

展具储藏（与展厅关系直接）：60 m^2。

工休间（展厅工作人员工作、休息）：60 m^2。

报告厅（包括进厅、器材、休息）：300 m^2（可单独对外开放）。

②收藏保管空间：约700 m^2。

办公（兼藏品登记、编目）：60 m^2。

藏品库：400 m^2（封闭式管理，注意安全性）。

工作间（修复、裱画、照相、复制等）：4~6间，240 m^2（要求房间可分隔使用）。

③技术、研究空间：约240 m^2。

研究创作室4~6间，要求房间可分隔使用，自然采光。

④行政办公空间：约150 m^2。

办公室：6间×15 m^2。

小型会议室：60 m^2。

⑤休闲服务空间：约300 m^2。

咖啡茶座、艺术品商店等

⑥其他空间：

值班室：10 m^2。

设备间：200 m^2（可分为若干间，分层设置）。

客用、货用电梯各一部。

其他门厅、交通、服务等空间面积根据设计需要自行确定。

4. 成果要求

图纸数量：不少于4张A1竖幅（597 mm×840 mm）。

图纸内容：

1）场地与环境：场地分析与说明，总平面（1：500），鸟瞰图。

2）空间生成与解析：空间概念生成分析，室内外空间构成的表达（轴测、剖透视、分解图等），空间氛围与效果的表达（人眼透视、序列局部透视等）。

3）技术图纸：各层平面，建筑沿街立面（不少于3道），剖面（不少于2道），比例1：200。

4）外观透视：模型，根据不同设计阶段的需求，制作不同深度的工作模型，最终表现模型比例另详。

1. Teaching objective

This course, themed in "space", intends to enable the students to learn architectural space organization skills and methods, and train them in terms of the operation and expression of space.

The problem of space is a basic problem of architecture. Based on the training and learning of complex spatial organization, this course focuses on the spatial order, and arranges the large space and small space, independent space and repetitive space, and distinguishes the public space from private space, serving space from service space, and open space from closed space. At the same time, it also fully considers people's behaviors and space perception in space, and tries to express specific meaning and atmosphere by space, thus forming a complete design.

2. Base and tasks

Nanjing, as an ancient capital of ten dynasties, has been a place with numerous celebrities. Therefore, it is planned to build a memorial of famous calligrapher & painter in the historical and cultural block of Changjiang Road, so as to promote the development of social and cultural undertakings. Each student should select a famous calligrapher & painter born in Nanjing, living in Nanjing for a long time, or closely related to Nanjing, analyze calligraphy and painting style or personal characters, excavate the memorial theme, and lead the organization and design of architectural space.

The base is located at the north side of the old building of Jiangsu Art Museum, namely the southwest corner of Beiting Lane and Shipopo Nunnery, with the land area of about 4,100 m^2.

The new building should have the following functions:
1) Show the life, story, works and influence of the calligrapher & painter;
2) Collect the materials, files, cultural relics, and works related to the calligrapher & painter;
3) Study the history, theory, creation etc. relevant to the calligrapher & painter;
4) Art popularization and social education functions, such as holding art lectures, seminars, salons, etc.

3. Space plan

1) Planning limitation
The total height of the building should not exceed 24 m, and the plot ratio should not exceed 1.2.
2) Space organization principle
Space organization should have clear characteristics, intention and concept, and should be able to meet the requirements of reasonable function, coordinated environment, and convenient circulation. Attention should be paid to the composition of space in different types and forms, the series organization of space, and the creation of space atmosphere.
3) Space type
Total building area: 4,500 m^2 (allowable error: 5%).

(The area ratio of each part below is for reference only, and each can adjust according to the study plan)
① Exhibition space: About 2,000 m^2.
Exhibition hall (or gallery): 1,500 m^2 (Several exhibition areas can be set, including temporary exhibition halls).
Exhibition appliance storage (directly related to the exhibition hall): 60 m^2.
Workplace (for working and resting by exhibition hall staff): 60 m^2.
Conference hall (including the entrance hall, equipment and resting area): 300 m^2 (Can be opened separately).
② Storage space: About 700 m^2
Office (collection registration and cataloging): 60 m^2.
Collection library: 400 m^2 (Closed management, safety should be focused).
Workplace (repairing, mounting, photographing, and copying): 4~6 rooms, 240 m^2 (They should be able to be separately used).
③ Technical research space: About 240 m^2.
4~6 research studios, with natural lighting, which should be able to be separately used.
④ Administrative office: About 150 m^2.
Office: 6 ones×15 m^2.
Small conference room: 60 m^2.
⑤ Leisure service space: About 300 m^2.
Cafe and teahouse, and art shop, etc..
⑥ Other space:
Duty room: 10 m^2.
Equipment room: 200 m^2 (can be divided into several rooms, set in layers).
One passenger elevator and one freight elevator.
The area of other halls, traffic and service space should be determined by the design needs.

4. Achievement requirements

Number of drawings: No less than four A1 vertical drawings (597 mm×840 mm).
Drawing content:
1) Site and environment: Site analysis and description, general plan (1: 500), aerial view.
2) Space generation and analysis: Space concept generation analysis, expression of indoor and outdoor space composition (axonometric, sectional perspective, and breakdown drawing), expression of space atmosphere and effect.(eye perspective, and sequence partial perspective)
3) Technical drawings: Plan of each floor, elevation along the street (no less than 3), section (no less than 2), scale 1: 200.
4) Appearance perspective:Model, develop working models of different depths according to the needs of different design stages, and the scale of the final presentation model should be determined separately.

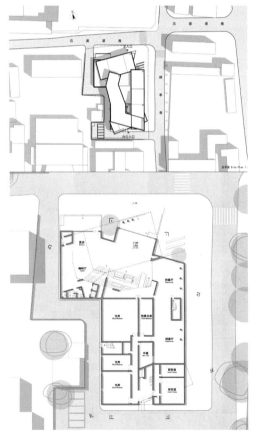
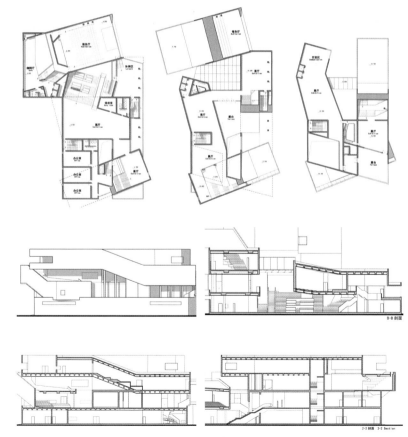

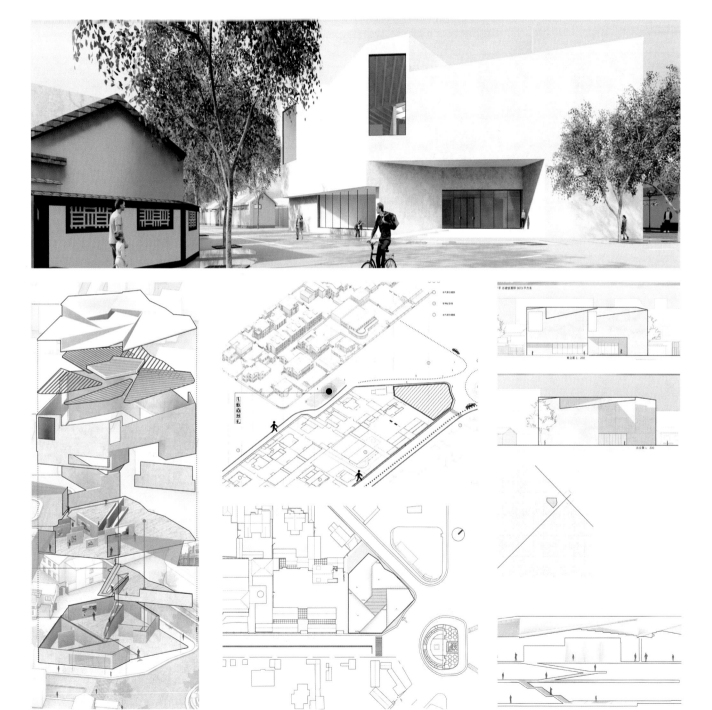

茶书吧设计
THE TEAROOM DESIGN
孟宪川

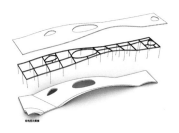

结构图示意图

1.教学目标

基本问题是建立起结构概念（基于图解静力学）与建筑设计的内在逻辑关系。

前提包括：1）较小尺度；2）坡屋顶意向；3）以结构为主，建筑功能较简单，防火分区基本不用划分。

力流概念图解对方案推敲的作用（分数占比15%）。

手工结构模型对方案推敲的作用（分数占比15%）。

查找相关知识并学习使用（分数占比5%）。

表现（分数占比5%）。

核心问题：结构如何激发建筑设计？

2.教学内容

建筑与结构的关系（最终通过图解表达）。

场地与结构的关系（场地回应策略）。

空间与结构的关系（结构对空间体验的影响）。

功能与结构的关系（结构对功能组织的影响）。

材料与结构的关系（材料断面尺寸，结构材料只能四选一：砖结构、钢筋混凝土、钢结构、钢木结构）。

构造与结构的关系（材料交接方法）。

图解静力学的辅助方法（最终通过图解表达）。

图解静力学概念与空间形态的关系（建筑学的基本问题）。

图解静力学图解对比与建筑设计方案推敲的关系（形式推敲的方法之一）。

竖向传力与水平传力的结构及形态解决措施（影响建筑学的重要问题）。

3.任务书

本项目位于南京城南老门东地区，现为一处沿街四进院落基地，基地占地约760 m²（约12 m×60 m）。

设定现状建筑已破败不可使用，但鉴于此处地块已形成商业旅游片区，决定在此处重新设计一文创产品体验店，使游客既可以看书休憩，又可以喝咖啡或饮茶交流。本项目强调对新建筑形态的探索（注意：不是传统建筑的重建）。设计要求为：1）建筑形态需适应周边传统建筑形式，即坡顶为主要设计意向；2）限高8 m（屋脊高度）；3）以结构力流的方式推敲和设计（本次设计中，学生需要绘制相关各类图解进行表达）；4）表达力流的结构模型+最终方案的结构构架模型作为设计辅助工具。用地红线内建筑全部拆除，不保留。

总建筑面积600 m²（±10%），建筑密度≤80%，总高不超过8 m，其中使用面积：

书吧空间：200 m²（包括展示、阅读体验、收银服务等空间）。

咖啡|茶书空间：200 m²（包括吧台、备餐、享用等空间）。

辅助用房：50 m²（包括存储、卫生间等空间）。

室外场地：150 m²（可与其他空间融合）。

4.成果要求

1）中期：（A0幅面2张，可黑白，允许线稿占位）+结构概念力流传递模型。

总平面图1:500。

平面图1:100。

立面图（沿街立面+开间入口立面）1:100。

透视图（沿街视角剖透视，至少反映出内部空间的结构、材料、人物活动等）。

鸟瞰图。

其他小透视。

分析图（图解至少包括：结构概念与形态的关系、形式推敲过程中结构的辅助方式、竖向力与水平力的解决策略，此部分占总分15%）。

技术经济指标（建筑面积、建筑密度、容积率、绿地率、建筑高度等）。

结构模型2个（表达力流传力的结构模型1:50和结构概念局部模型1:20，此部分占总分15%）。

工作手册（每周成果简要呈现）。

2）期终：A0幅面2张+结构模型2个+工作手册1份。

总平面图1:500。

平面图1:100。

立面图（沿街立面+开间入口立面）1:100。

透视图。

鸟瞰图。

分析图（图解至少包括：结构概念与形态的关系、形式推敲过程中结构的辅助方式、竖向力与水平力的解决策略，此部分占总分15%）。

技术经济指标（建筑面积、建筑密度、容积率、绿地率、建筑高度等）。

结构模型2个（表达力流传力的结构模型1:100+呈现方案构件的结构模型1:50，此部分占总分15%）。

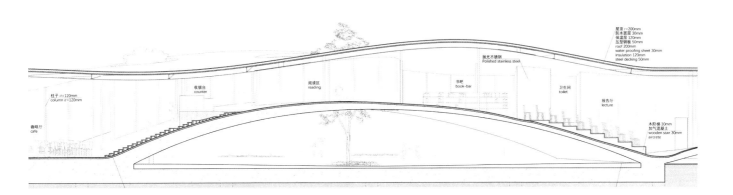

1. Teaching objective
The basic problem is how to establish an internal logical relationship between the structural concept (based on graphic statics) and architectural design.
The premises include: 1) Smaller scale; 2) Sloping roof intention; 3) Focus on the design of structure with relative simple functions, and fire compartments are basically not divided.The effect of the force flow concept diagram on scheme deliberation (proportion: 15%).
The effect of manual structure model on scheme deliberation (proportion: 15%).
Seek the relevant knowledge and learn to use (proportion: 5%).
Performance (proportion: 5%).
Core problem: How does structure stimulate architectural design?

2. Teaching contents
The relationship between architecture and structure (finally expressed by a diagram).
The relationship between site and structure (site response strategy).
The relationship between space and structure (the influence of structure on spatial experience).
The relationship between function and structure (the influence of structure on function organization).
The relationship between materials and structure (cross-section size, structural material can only be one of the four: Brick structure, reinforced concrete, steel structure, steel-wood structure).
The relationship between construction and structure (material connection method).
Auxiliary method of graphic statics (finally expressed by a diagram).
The relationship between the concept of structuring concept and space form based on graphic statics(basic problem of architecture).
The relationship between graphic statics graphic comparison and architectural design scheme deliberation(One of the methods of formal deliberation).
The structural and morphological solutions for vertical and horizontal force transmission (an important problem affecting the architecture).

3. Design brief
The project is located in Laomendong district in south of Nanjing city. It is now a 4 rows courtyard base along the street. The base covers an area of about 760 m² (about 12 m × 60 m).
Assume that the architecture is dilapidated and cannot be used, but consider its location in a commercial tourism area, it is determined to be re-designed as a cultural and creative product experience store, for visitors to read and rest, or communicate with each other while drinking coffee or tea. This project emphasizes the exploration of a new architectural form (note: not the reconstruction of a traditional building).
Design requirements: 1) The architectural form should adapt to the surrounding traditional architecture which means that the design should focus on the roof; 2) Height limit: 8 m (the height of ridge); 3) Deliberation and design should be performed by means of structural force flow (in this design, the students should draw various diagrams for expression); 4) The structural model expressing the force flow + the structural framework model of the final scheme should be taken as the auxiliary means of design. All buildings within the range of the red line should be demolished.
Total building area: 600 m² (±10%), building density ≤80%, total height ≤8m, and the usable area is shown below:
Book bar space: 200 m² (including the space for display, reading experience and cashier service).
Cafe|tea bar space: 200 m² (including the space for bar counter, meal preparation area, and enjoyment area).
Auxiliary room: 50 m² (including storage space, and toilet).
Outdoor site: 150 m² (it can be integrated with other space).

4. Achievement requirements
1) Mid-term: (2 A0 drawings, black and white, allowing handmade placeholder) + structural conceptual force flow transmission model.
General layout (1 : 500).
Plan (1 : 100).
Elevation (Elevation along the street + Entrance elevation) (1 : 100).
Perspective (sectional perspective along the street, reflecting the structure, materials, and activities of the internal space at least).
Aerial view.
Other small perspectives.
Analysis chart (it must include the relationship between structural concept and form, the structure auxiliary method in the process of form deliberation, vertical force and horizontal force resolution strategy; this part accounts for 15% of the total score).
Technical and economic indicators (building area, building density, plot ratio, greening rate and architecture height).
2 structural models [the structural model expressing force flow (1 : 50) and partial model of structural concept (1 : 20); this part accounts for 15% of the total score].
Work manual (brief presentation of weekly achievement).
2) End of term: 2 A0 drawings + 2 structural models + 1 work manual.
General layout (1 : 500); Plan (1 : 100); Elevation (elevation along the street + entrance elevation) (1 : 100); Perspective; Aerial view; Analysis chart (it must include the relationship between structural concept and form, the structure auxiliary method in the process of form deliberation, vertical force and horizontal force resolution strategy; this part accounts for 15% of the total score).
Technical and economic indicators (building area, building density, plot ratio, greening rate, and architecture height).
2 structural models [the structural model expressing force flow (1 : 100) and structural model presenting components (1 : 50); this part accounts for 15% of the total score].

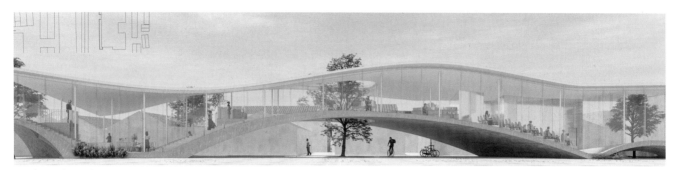

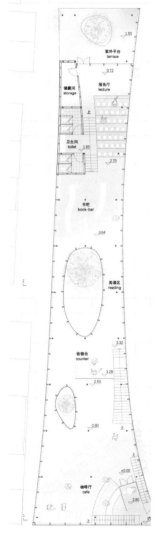

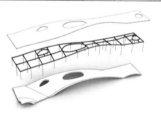

设计策略一：拱的荷载分布　　　　　　　设计策略二：拱的最高点位置

设计策略三：拱的壁厚

设计策略四：拱高

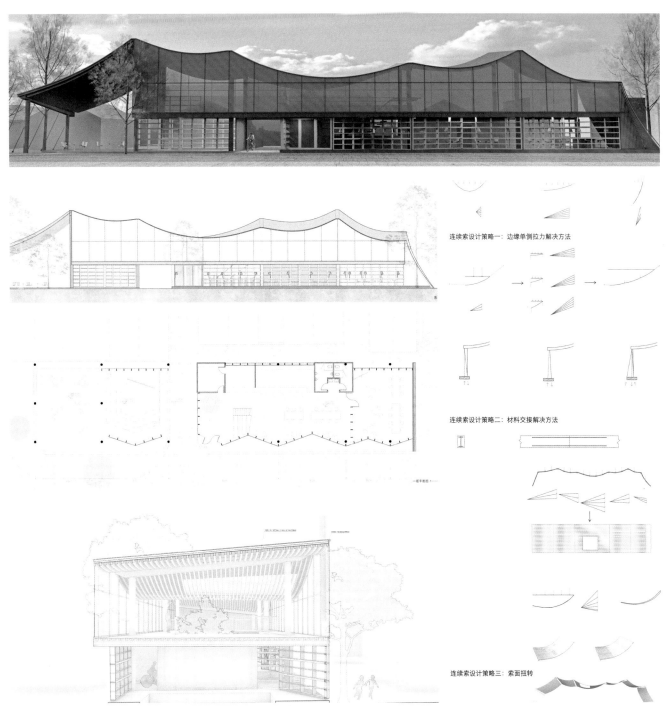

建筑设计（五）ARCHITECTURAL DESIGN 5

大学生健身中心改扩建设计
RENOVATION AND EXTENSION DESIGN OF COLLEGE STUDENT FITNESS CENTER

傅筱 王铠 钟华颖

1. 课程缘起

大学生健身中心改扩建设计是南京大学建筑与城市规划学院本科新开设的一个三年级设计课程。在南京大学过去的设计教学中一直缺少一个专项训练中大跨结构的设计课程，而是将其融入三年级商业综合体课程设计中。学生既要解决商业综合体的复杂城市关系和复合功能空间，又要研究其中的大跨度设计，其难度可想而知。大学生健身中心改扩建设计课程开设的初衷十分明确，即是将中大跨空间从商业综合体中分离出来进行分项训练。

2. 选题思考：弱规模，强认知

在中大跨度建筑设计训练方面国内许多建筑院校做出了值得学习的探索，它们多数是以大型体育馆、演艺中心或者文化中心为授课载体，以结构选型为基础，结合设备训练学生对大型建筑场馆的综合设计能力。有的院校将结构独立成8周专项训练，然后再结合进8周综合场馆设计中。这些训练通常是放置在四年级或者毕业设计，不少院校设计周期长达16周，让学生得到较为充分的训练。对于南京大学"2+2+2"的建筑教育模式而言，是不适宜进行长周期训练的。在短短的8周教学时间内，让学生建立认知和培养能力是关键。教学团队意识到高校课程设计并非实际工程训练，与漫长的职业生涯相比，学生的认知能力比实际操作能力更为重要，设计的规模和技术的复杂性并不一定能提升他们对设计本质的理解。因此，结合南京大学自身的教学特征，教学团队在选题上做了如下思考：

将结构训练的重点放在结构与形态、空间的关联性上。在以前的课程设计中学生较为熟悉的是框架、砖混结构，虽然每一次课程设计均与结构不可分割，但是结构被外围护材料包裹，在建筑表达上始终处于被动"配合"的状态。对于有一定跨度的空间，容易让学生建立一种"主动"运用结构的意识。为了达到这个训练目标，教学团队认为跨度不宜太大，以30 m左右为宜，结构选型余地较大，结构与空间的配合受到技术条件的制约相对较小，以利于学生充分理解空间语言与结构的关联。

强调建筑设计训练的综合性，结构只是一个重要的要素，即不过分放大结构的作用，在将结构作为空间生成的一种推动力的同时要求学生综合考虑场地、使用、空间感受、采光通风等基本要素。实际上通过短短8周教学就希望学生能够达到实际操作层面的综合性是不现实的，教学的目标是让学生建立起综合性的认知，所以题目设置宜简化使用功能，明确物理性能要求，给定设备和辅助空间要求，将琐碎的知识点化为知识模块给定学生，学生更多的是学会模块间的组织，而不需要确切了解其中的每一个技术细节。技术细节日新月异，将随着学生今后的职业生涯而不断增长，在高校教授的许多的技术细节或者教师的所谓工程经验，也许到学生毕业时就已经面临淘汰。所以在高校应教授"认知"，进而教授由正确认知带来的技术方法，最终升华为学生的设计哲学。因此，在具体的教学中，结构选型是指定的，但鼓励学生结合性能需求进行合理改变；采光通风要求是明确的，但鼓励结合人的需求进行设计；设备和辅助空间是给定的，但要求学生学会在空间上进行合理布置，理解只有设备和辅助空间布置妥当才能创造使用空间的价值。

3. 教学方法：从"一对一"到"一对多"

本科建筑设计教学方法通常采用教师一对一改图模式，这种模式的优点是教师易于将设计建议传递给学生，当学生不理解时，还可以方便地用草图演示给学生，这适合于手绘时代的交流，甚至教师帅气的草图也是激励学生进步的一种手段。其缺点是学生的问题和教师的建议均局限在一对一的情景中，不能让更多的学生受益，如果教师发现共性问题须临时召集学生再行讲解，其时效性、生动性都大打折扣。鉴于此，我们采用了"一对多"的改图模式，教师和学生全部围坐在投影仪前面，由每个学生讲解自己的设计，教师做点评，并同时鼓励其他学生发表点评意见，当教师认为需要辅助草图说明问题时，教师采用 Pad 绘制草图，并让每个学生都看见。

从教学效果看，一对多的改图模式其优点是明显的，首先这样的改图方式具有较好的课堂氛围，学生面对大屏幕讲解自己的设计增强了课堂的仪式感，仪式感的背后是对学习的敬畏之心，同时因为教师不是站在高高的讲台上，又具有平等讨论的轻松气氛。从教学效果上看，一方面让学生既体验到当众讲解的压力，也体验到因出色的设计呈现带来的喜悦，因而无形中激发了他们用功的积极性；另一方面有利于教师发现学生的共性问题，避免同一问题反复讲却反复犯的教学通病，大大提高了课堂效率。

1. The background of course

The renovation and extension design of college student fitness center is a new third-year design course of the School of Architecture and Planning, Nanjing University. In the past design teaching of Nanjing University, there was always short of a special training course of long-span structure design, which was integrated into the course design of the third-year commercial complex. The students should not only solve the complex urban relationship and composite function space of commercial complex, but also study the long-span design, the difficulty of the course is not hard to imagine. The original intention of setting the course of renovation and extension design of college student fitness center is very clear, that is, to separate the long-span space from the commercial complex for itemized training.

   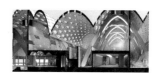
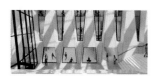   

2.Topic thinking: Weaken scale, strengthen cognition

In terms of architectural design training of long-span, many architectural colleges in China have made the exploration which is worth learning. Most of them take medium and large gymnasiums, performance centers or cultural centers as teaching carriers, and train students' comprehensive design ability for large construction venues based on structure selection and combined with equipment. Some colleges separate the structure into 8 weeks of special training, and then combine it into 8 weeks of comprehensive venue design. These trainings are usually placed in the fourth grade or graduation project, the design cycle of many colleges is up to 16 weeks, so that the students could get adequate trainings. As for the "2+2+2" architectural education mode of Nanjing University, it is not suitable for long-periodic training. In just eight weeks of teaching, the key is to help students to establish some kind of cognition and develop some kind of ability. The teaching team realized that college curriculum design was not practical engineering training, the students' cognitive ability was more important than their practical operating ability compared with their long career, and the scale and complexity of design did not necessarily improve their understanding of the nature of design. Therefore, combined with Nanjing University's own teaching characteristics, the teaching team made the following considerations on the topic selection:

The emphasis of structural training is on the relationship between structure, form and space. In the previous course design, students are familiar with the frame and brick-concrete structure. Although each course design is inseparable from the structure, the structure is wrapped by the peripheral materials and is always in a passive state of "cooperation" in architectural expression. For a space with a certain span, it is easy for students to establish a sense of "actively" using the structure. In order to achieve this training goal, the teaching team believes that the span should not be too large, about 30 m is appropriate, and there is a large scope of structure selection, and the coordination between structure and space is relatively limited by technical conditions, so as to facilitate students to fully understand the correlation between spatial language and structure.

The emphasis on the integration of architectural design training, structure is only an important element. That is to say, the function of the structure is not exaggerated. While taking the structure as a driving force for the generation of space, students are required to comprehensively consider the basic elements of site, use, space perception, lighting and ventilation. In fact, it is unrealistic to expect students to achieve practical comprehensiveness in just eight weeks' teaching. The teaching goal is to enable students to establish a comprehensive cognition, so the topic setting should simplify the use function, clarify the physical performance requirements, set equipment and auxiliary space requirements, and turn trivial knowledge points into knowledge modules for students. Then, students should learn more about the organization of modules, and do not need to know each of these technical details. Technical details will continue to grow with the students' future career, which in today is really changing with each passing day. Many technical details or teachers' so-called engineering experience are taught in colleges and universities, which may be obsolete by the time they graduate. Therefore, "cognition" should be taught in colleges and universities, and then the technical methods brought by correct cognition should be taught, which finally developed design philosophy of students. Hence, in specific teaching, structure selection is specified, but students are encouraged to make reasonable changes based on performance requirements; The requirements for lighting and ventilation are clear, but students are encouraged to design according to human needs; Equipment and auxiliary space are given, but students are required to learn to arrange the space reasonably, and understand that only when the equipment and auxiliary space are properly arranged can the value of using space be created.

3.Teaching method: From "one-to-one" to "one-to-many"

The undergraduate architectural design teaching method usually adopts the mode of one-to-one figure modification by teachers, the advantage of this model is that teachers can easily pass design suggestions to students, and when students don't understand, they can also easily demonstrate to students with sketches, which is suitable for communication in the hand-drawn era. Even the handsome sketches of teachers are means to encourage students' progress. And the disadvantage is that students' problems and teachers' suggestions are limited to one-to-one situation, which cannot benefit more students. If teachers find common problems and need to call on students to explain them temporarily, the timeliness and vividness will be greatly reduced. In view of this, we adopted "one to many" figure modification mode, all teachers and students sit in front of the projector, then, each student explains his own design, the teacher makes comments, and encourages other students to give comments. When the teacher thinks it is necessary to help the sketch to explain the problem, the teacher uses Pad to make the sketch, which can be seen by every student.

From the perspective of teaching effect, the advantages of one-to-many figure modification mode are obvious. First of all, this kind of figure modification way has a good classroom atmosphere. Students explain their design in front of the big screen, which enhances the sense of ceremony in class. Behind the sense of ceremony is the respect of learning, and because the teacher is not standing on the high platform, there is a relaxed atmosphere of equal discussion. From the perspective of teaching effect, on the one hand students not only experience the pressure of public explanation, but also experience the joy brought by excellent design presentation, thus stimulating their enthusiasm to work hard; On the other hand, it is beneficial for teachers to find out the common problems of students and avoid the common teaching mistakes of repeating the same problem, which greatly improves the classroom efficiency.

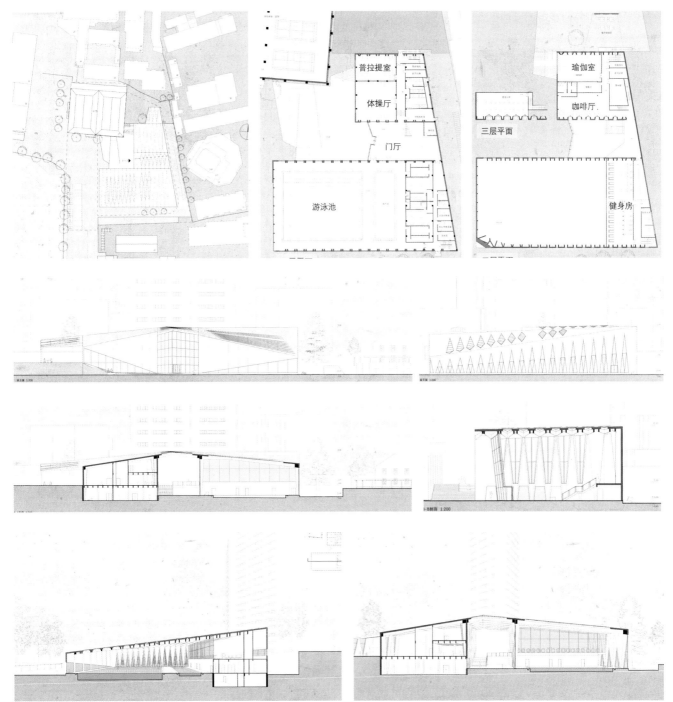

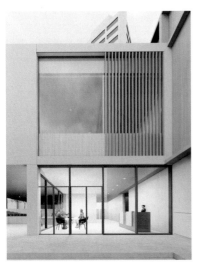

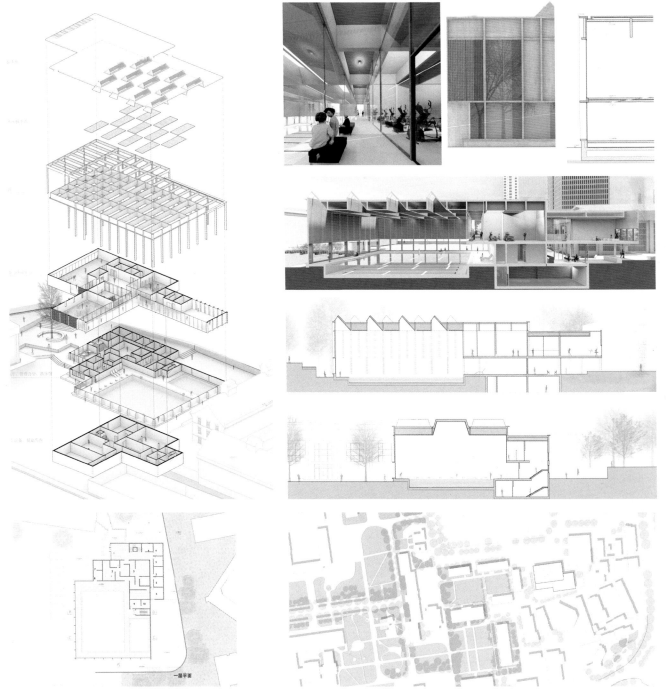

一层平面

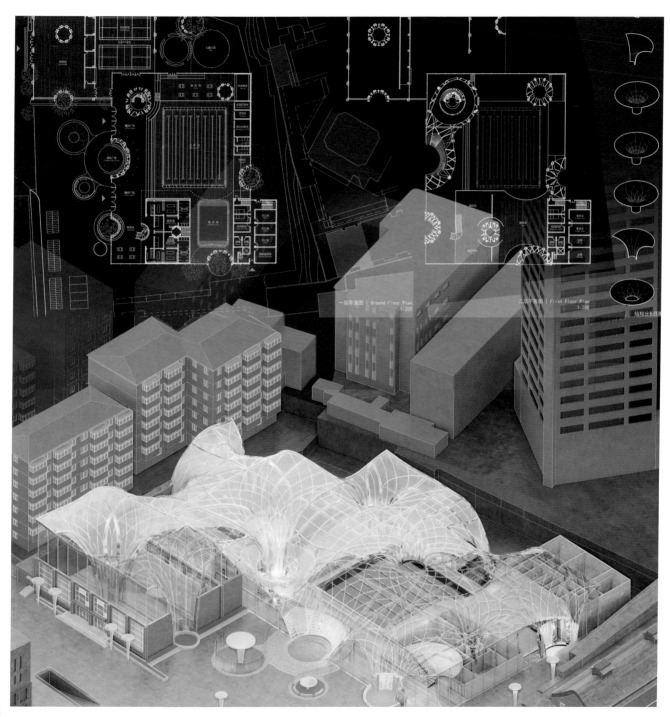

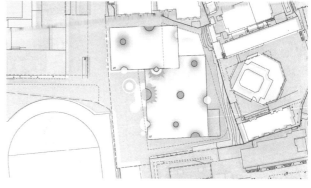

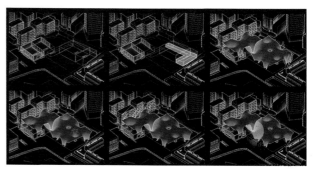

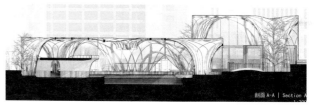

剖面 A-A | Section A

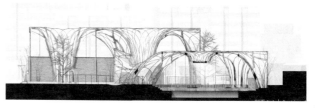

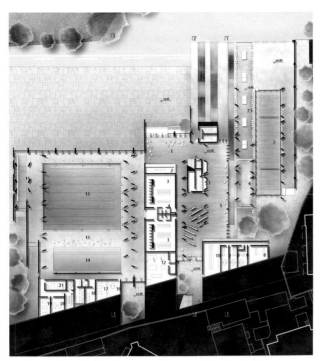

社区文化艺术中心设计
COMMUNITY CULTURE AND ENTERTAINMENT CENTER DESIGN

张雷 钟华颖 王铠

1. 概况
本项目拟在百子亭风貌区基地处新建社区文化中心，总建筑面积约8000 m²，项目不仅为周边居民文化基础设施服务，同时也期望成为复兴老城的街区活力的文化地标。根据基地条件、功能使用进行建筑和场地设计。总用地详见附图，基地用地面积4600 m²。民国时期，百子亭一带属于高级住宅区，在位置上紧邻作为文教区的鼓楼，以及作为市级行政区的傅厚岗地区。凭借区域上的优势与政府的扶持，百子亭一带自1930年代开始，逐渐成为当时文化精英、社会名流与政府要员的聚居之地。众多受邀前往南京创建其事业的学者、文人都在此购买土地，并建造出了一幢幢"和而不同"的新式住宅。这些建筑既是近代南京城市肌理中的现代图景，也是当时中国有为之士们实践其梦想的舞台，还是中国近现代建筑史中不可忽视的华美段落。

2. 基地条件：现状
根据《南京历史文化名城保护规划》（2010版），百子亭风貌区被列入"保护名录"，风貌区内现有市级文物保护单位3处，为桂永清公馆旧址、徐悲鸿故居及傅抱石故居，不可移动文物8处，历史建筑1处。

3. 设计内容
1）演艺中心
包含400座小剧场，乙级。台口尺寸12 m×7 m。根据设计的等级确定前厅、休息厅、观众厅、舞台等面积。观众厅主要为小型话剧及戏剧表演而设置。按60~80人化妆布置化妆室及服装道具室，并设2~4间小化妆室。要求有合理的舞台及后台布置，应设有排练厅、休息室、候场区以及道具存放间等设施，其余根据需要自定。

2）文化中心
定位于E级综合性文化站，包括公共图书阅览室、电子阅览室、多功能厅、排练厅、辅导培训、书画创作等功能室（不少于8个且每个功能室面积应不低于30 m²）。

3）配套商业
包含社区商业以及小型文创主题商业单元。其中社区商业为不小于200 m²超市一处，文创主题商业单元面积为60~200 m²。

4）其他
变电间、配电间、空调机房、售票、办公、厕所等服务设施根据相关设计规范确定，各个功能区可单独设置，也可统一考虑。地上不考虑机动车停车配建，在街区地下统一解决，但需要根据建筑功能面积计算车位数量。

4. 教学成果
每人不少于4张A1图纸，图纸内容包括：
1）城市与环境：总平面图1:500，总体鸟瞰图、轴测图。
2）空间基本表达：平、立、剖面图1:200~1:400（附比例尺）。
3）空间解析与表现：概念分析图、空间构成分析图、轴测分析图、剖透视（不少于2张，必须包含大空间、公共空间的剖透视）、室内外人眼透视若干。
4）手工模型：每个指导教师组内各做一个1:500总图体量模型，每位学生做一个1:500的概念体块模型。

5. 教学进程
本次设计课程共8周。
第一周：授课1学时、调研场地及案例、制作场地模型（SU模型+实体模型1:500）相应的案例资料收集。
第二周：学生收集案例汇报、初步概念方案讨论（包含体块与场地关系布局、内部空间基本布局）。
第三周：概念深化，完善初步建筑功能布局和空间形态方案（包括基本空间单元及其组合），制作空间形态模型。
第四周：方案定稿，明确空间表皮、平面功能、街区环境模式。
第五周：方案深化Ⅰ：空间表皮、平面功能、街区环境深化。
第六周：方案深化Ⅱ：细化表皮处理，剧场空间及其他重要公共空间设计。
第七周：方案表达：完成平、立、剖绘制，完善SU设计模型。
第八周：制图、排版。

1. Overview
This project plans to build a new community cultural center at the base in the Baiziting Scenic Area, with the total building area of about 8000 m². It will serve surrounding residents as the cultural infrastructure, and will also become a cultural landmark for revitalizing the neighborhood of the old town. The architectural and site design should be carried out according to base conditions and functions. The overall land use is shown in the attached drawing, and the base covers an area of 4600 m². During the period of the Republic of China, Baiziting is an exclusive residential district, which was next to the Gulou District (the cultural and educational district) and Fuhougang District (the municipal administrative district). With the regional advantages and

 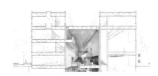  

government support, Baiziting gradually became a settlement of cultural elites, celebrities and government officials in the 1930s. The scholars and literati who were invited to Nanjing bought land and built their own new-type houses featured by "harmony in diversity". These buildings were the modern factors of Nanjing and the stages for those scholars and literati to achieve their dreams; in addition, they also formed a gorgeous passage that cannot be ignored in the history of modern Chinese architecture.

2. Base conditions: Current status
According to *Nanjing Historic and Cultural City Conservation Plan* (2010), Baiziting Scenic Area was listed in the "protection catalogue". At present, there are 3 municipality protected historic sites (Former Residence of Gui Yongqing, Former Residence of Xu Beihong, and Former Residence of Fu Baoshi), 8 immovable cultural relics and 1 historical building.

3. Design contents
1) Performance center
It contains 400 small theaters (Class B), with the entablature size of 12 m×7 m. The area of the front hall, lounge, auditorium, and stage should be determined according to the design level. The auditorium is mainly set for small drama performances. The dressing room and costume prop room should be arranged for 60~80 people, and 2~4 small dressing rooms should be set. There should be reasonable stage and backstage arrangements, as well as rehearsal halls, lounges, waiting areas and prop storage rooms; and other facilities should be determined as required.

2) Cultural center
It is a district-level comprehensive cultural station, which consists of the public reading room, electronic reading room, function hall, rehearsal hall, training room, and calligraphy and painting creation room (at least 8, and each should have an area of 30 m^2).

3) Supporting business
It contains community business and small cultural and creative theme business units. There is one supermarket of no less than 200 m^2, and the cultural and creative commercial units cover an area of 60~200 m^2.

4) Others
The substation, power distribution room, air-conditioning room, ticket office, office, and toilet should be determined according to the relevant design specifications. The functional areas can be set individually or considered together. There would be no ground parking lots, but underground ones, whose number should be calculated according to the functional area of the building.

4. Teaching achievements
Each one should have at least four A1 drawings, and the content of the drawings include:
1) City and environment: General layout (1 : 500), overall aerial view, and axonometric drawing.
2) Basic expression of space: Plane, elevation and section (1 : 200~1 : 400) (with scale).
3) Spatial analysis and performance: Conceptual analysis chart, space composition analysis chart, axonometric analysis diagram, sectional perspective (no less than 2 drawings, including sectional perspective of large space and public space), and several perspectives indoors and outdoors.
4) Manual model: In each instructor group, a 1 : 500 overall volume model should be established, and each student should prepare a 1 : 500 conceptual volume model.

5. Teaching process
This design course lasts for 8 weeks
Week 1: Teach for 1 class hour, investigate the site and case, and collect case materials of site model (SU mode I + entity model 1 : 500).
Week 2: The students should collect case reports, and discuss preliminary concept plans (including the layout of the relationship between the block and site, and basic layout of the internal space).
Week 3: Perform concept deepening, and improve preliminary architectural function layout and spatial form plan (including basic space units and their combinations), and prepare space form models.
Week 4: Finalize the scheme, clarify the space surface, plane functions, and block environment model.
Week 5: Scheme deepening Ⅰ: Deepen the space surface, plane function, and block environment.
Week 6: Scheme deepening Ⅱ: Refine the design of skin treatment, theater space and other important public space.
Week 7: Scheme expression: Complete the plane, elevation and section, and improve the SU design model.
Week 8: Charting, typesetting.

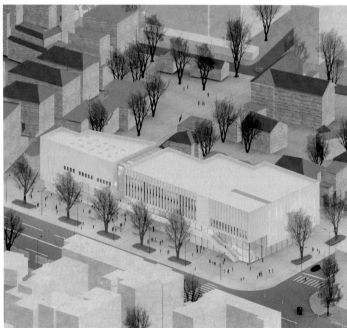

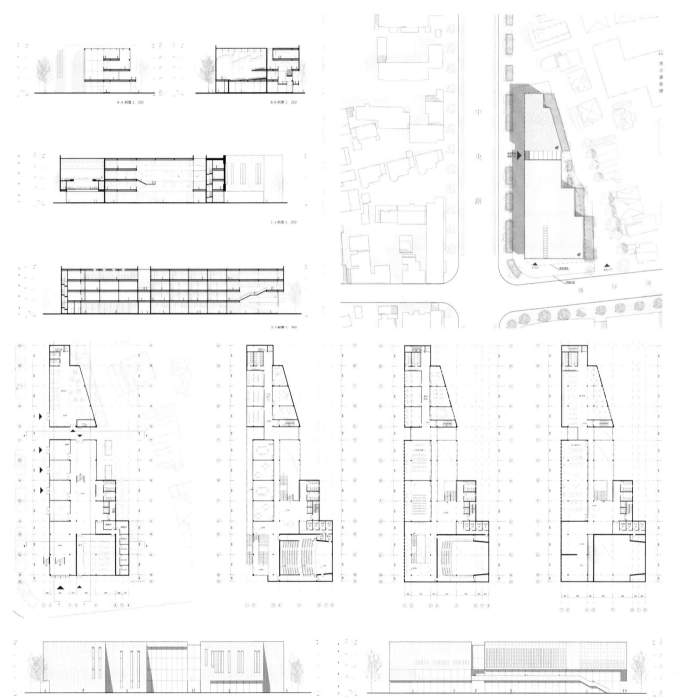

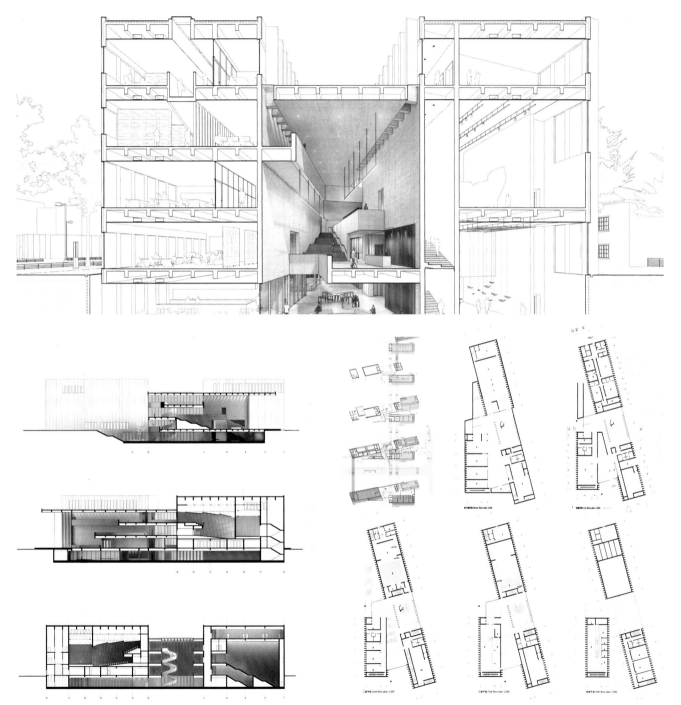

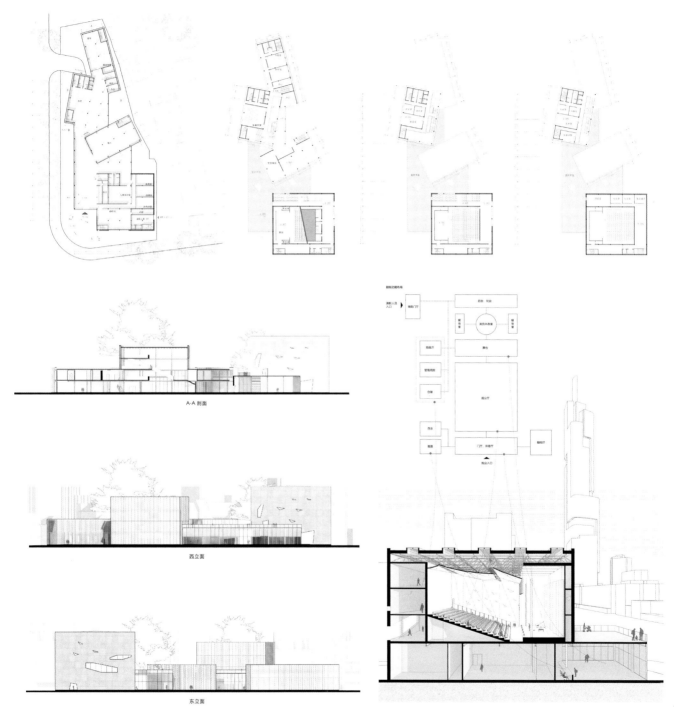

建筑设计（七）ARCHITECTURAL DESIGN 7

高层办公楼设计
DESIGN OF HIGH-RISE OFFICE BUILDING

吉国华 胡友培 尹航

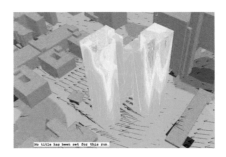

1. 教学目标

生态性能驱动的办公建筑设计涉及城市、空间、形体、环境、能耗、结构、设备、材料、消防等方面内容，是一项较复杂与综合的任务。有效的空间组织、适应性形体、交互性表皮以及性能化构造设计等策略，对建筑室内外环境的生态性能起着决定性的作用。本课题教学重点和目标是帮助学生理解、消化以上涉及各方面知识，提高综合运用并创造性解决问题的技能，学习并运用生态性能模拟分析软件，以生态性能驱动建筑设计。

2. 项目设计条件与要求

1）经济技术指标与场地
用地面积4520 m²，地上总建筑面积≥35000 m²，建筑限高≤100 m。

2）功能要求

办公：设计应兼顾各种办公空间形式。

会议：须设置400人报告厅一个，200人报告厅2个，100人报告厅4个，其他各种会议形式的中小型会议室若干，以及咖啡茶室、休息厅、服务用房等。

机动车交通：机动车交通独立设置，人车分离。场地交通流线须结合现状周边情况统一考虑。地下部分为车库和设备用房。

3）相关规范

《民用建筑设计通则》 GB 50352—2005
《无障碍设计规范》 GB 50763—2012
《办公建筑设计规范》 JGJ/T 67—2019
《汽车库建筑设计规范》 JGJ 100—2015
《建筑设计防火规范》 GB 50016—2014
《汽车库、修车库、停车场设计防火规范》 GB 50067—2014

4）其他

用地红线及建筑退让线详见总平面图。汽车库和自行车库的配置应满足《南京市建筑物配建停车设施设置标准与准则》（2019）的要求。

底层的架空层面积不计入建筑总面积。

3. 教学进程（8周）

第一周：场地调研与分析、案例研究、初步概念。
第二周：概念深化。
第三、四周：总平面设计（草图、1∶500草模）。
第五、六、七周：平面、立面与细部深化设计。
第八周：成果表达与制作。

4. 成果要求

建筑总平面图（1∶500）。
建筑平、立、剖面图（1∶200）。
建筑大样图（大等于1∶20）。
建筑表现图若干。
建筑形体研究模型（1∶500）。
建筑模型（1∶200/1∶300）。

1. Teaching objective

The design of office buildings driven by eco-performance involves the aspects of city, space, form, environment, energy consumption, structure, equipment, materials, and fire protection etc.. It is a complex and comprehensive task. The strategies such as effective spatial organization, adaptive shape, interactive surface, and performance-based structural design play a decisive role in ecological performance of indoor and outdoor environment. This course intends to help the students to understand and digest the knowledge of various aspects, improve comprehensive application and creative problem-solving skills, learn and use the ecological performance simulation analysis software, and drive architectural design with ecological performance.

2. Project design conditions and requirements

1) Economic and technical indicators and site

Land area: 4520 m², the total building area above ground ⩾ 35 000 m², height limit ⩽ 100 m.

2) Functional requirements

Office: The design should take into account various forms of office space.

Meeting: There should be a 400-person conference hall, two 200-person conference halls, and four 100-person conference halls. There should also be several small and medium-sized conference rooms in various forms, and the cafe/tea bar, lounge, and service room.

Vehicle traffic: The vehicle traffic should be set independently, with separation between men and vehicle. Site traffic flow should be considered in combination with the surrounding situations. The underground part consists of garage and equipment room.

3) The relevant specifications

Code for Design of Civil Buildings (GB 50352—2005)

Code for Accessibility Design (GB 50763-2012)

Design Code for Office Buildings (JGJ/T 67—2019)

Code for Design of Garage Buildings (JGJ 100—2015)

Code for Fire Protection Design of Buildings (GB 50016—2014)

Code for Fire Protection Design of Garage, Motor-repair-shop and Parking-area (GB 50067—2014)

4) Others

See the red line of land use and building set-back line in the general layout. The configuration of garage and bicycle garage should meet the requirements of the *Standards and Guidelines for Setting Parking Facilities for Buildings in Nanjing* (2019).

The area of the raised floor is not included in the total building area.

3. Teaching process (8 weeks)

Week 1: site investigation and analysis, case research, initial concept.

Week 2: concept deepening.

Week 3 & 4: general plan design (sketch, draft model - 1∶500).

Week 5, 6 & 7: plane, elevation and detail deepening design.

Week 8: Students' works expression and preparation.

4. Achievement requirements

General layout (1∶500).

Plane, elevation, and section (1∶200).

Detail drawing (⩾1∶20).

Several architectural performance drawings.

Building form research model (1∶500).

Building model (1∶200/1∶300).

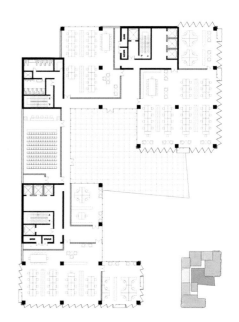

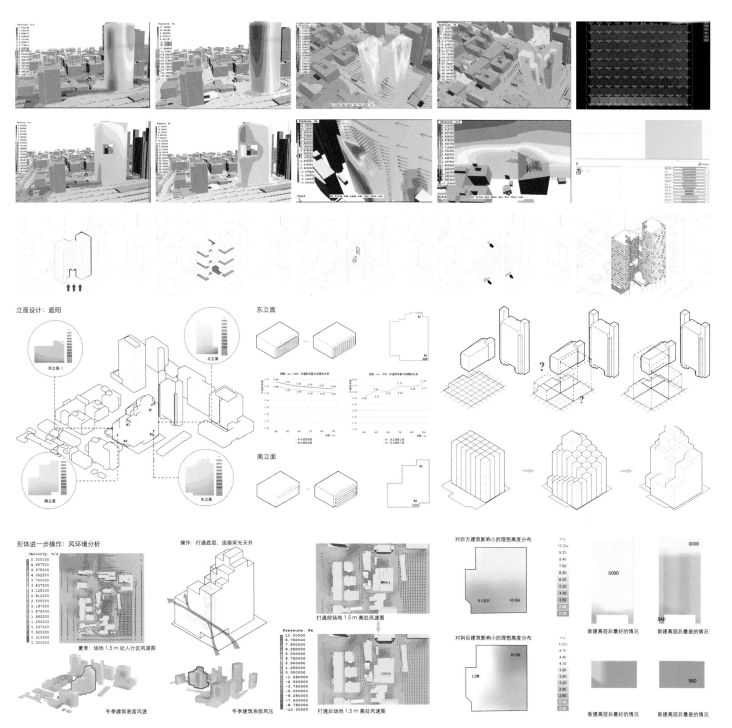

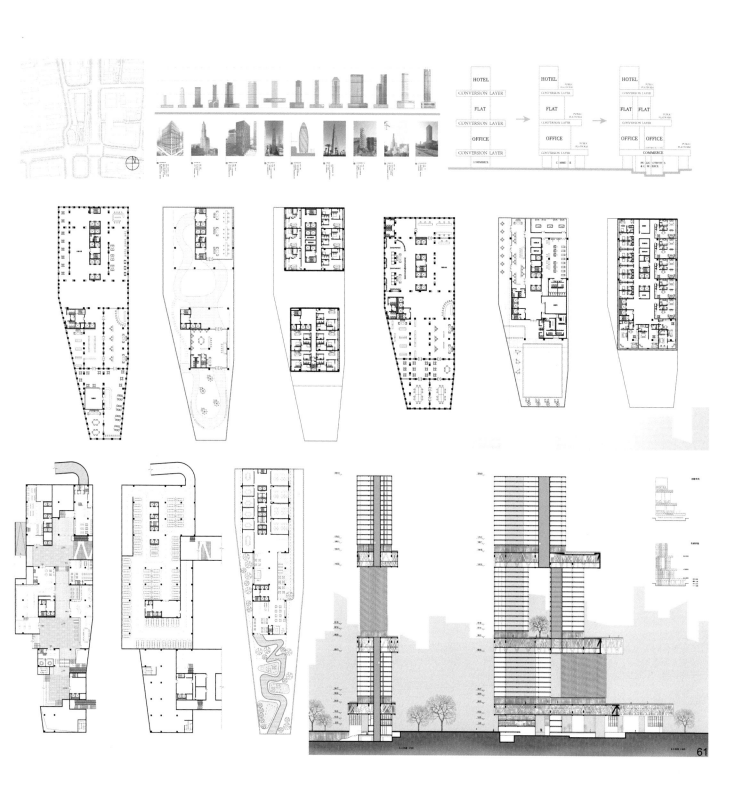

建筑设计（八）ARCHITECTURAL DESIGN 8

城 市 设 计
URBAN DESIGN

童滋雨 唐莲 尤伟

1.教学目标
1）基于真实地块的设计训练，认知高密度城市空间形态的真正含义，了解城市建筑角色和城市物质空间的本质和效能。
2）理解城市设计操作对象与设计内容，建立城市空间感知质量与城市建筑组合之间的关系，掌握城市设计方法。
3）在设计研究的过程中，进一步深化空间设计的技能、方法与绘图能力。

2.教学内容
我国沿海城市面临城市空间发展和土地资源匮乏的矛盾，土地资源立体化、高密度地使用变得极其重要，城市设计的角色正是在高密度情况下探索高品质空间塑造的可能性。为此，本次设计以南京市河西新城中心商务区的地块为操作对象，通过加大密度、重新分配功能与重组交通的方式提高土地使用效率，强调以空间中的感知质量比如围合感、可达性、可视性等为目标导向，塑造优质的城市空间。

3.关键词
城市肌理形态、城市空间密度、城市空间感知质量。

4.成果要求
1）总体设计（小组）2 A0。
2）深化设计（个人）2 A0。
3）模型（小组+个人）。

1.Teaching objective
1) Based on the design training of real plots, recognize the true meaning of high-density urban space, and understand the essence and efficacy of urban architecture role and urban physical space.
2) Understand the operation objects and design contents of urban design, establish the relationship between the perceived quality of urban space and urban building combination, and master urban design methods.
3) In the process of design research, further deepen the space design skills, methods and drawing competence.

2.Teaching contents
The coastal cities in China face the contradiction between urban spatial development and lack of land resources, so it is extremely important to realize three-dimensional and high-density use of land resources; the role of urban design is to explore the possibility of high-quality space shaping under high-density conditions. Therefore, this design takes the plot in the central business district of Hexi New Town, Nanjing City, so as to improve the efficiency of land use, emphasize the objective orientation of perception quality in space, such as the sense of enclosure, accessibility, and visibility through increasing density, redistributing functions and reorganizing traffic, thus creating the high-quality urban space.

3.Key words
Urban texture form, urban spatial density, urban space perception quality

4.Achievement requirements
1) Overall design (group) 2 A0.
2) Deepened design (individual) 2 A0.
3) Model (group+individual).

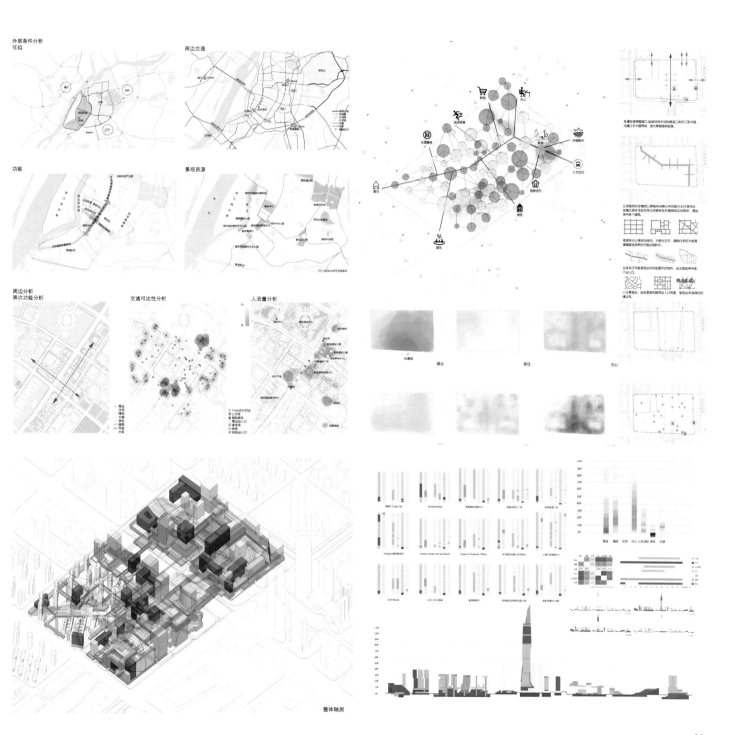

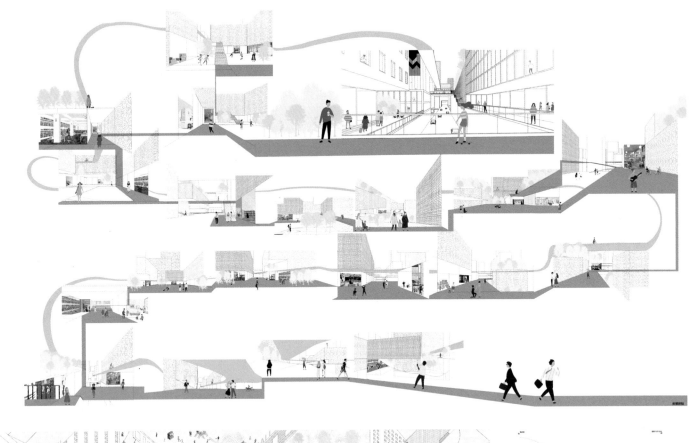
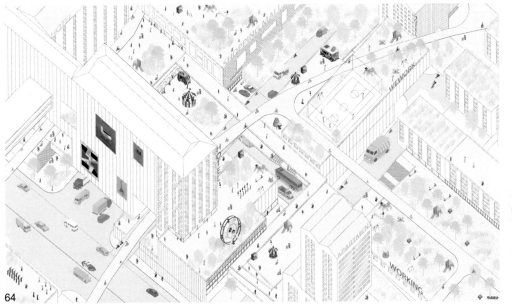
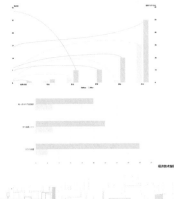
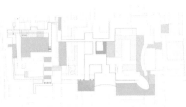

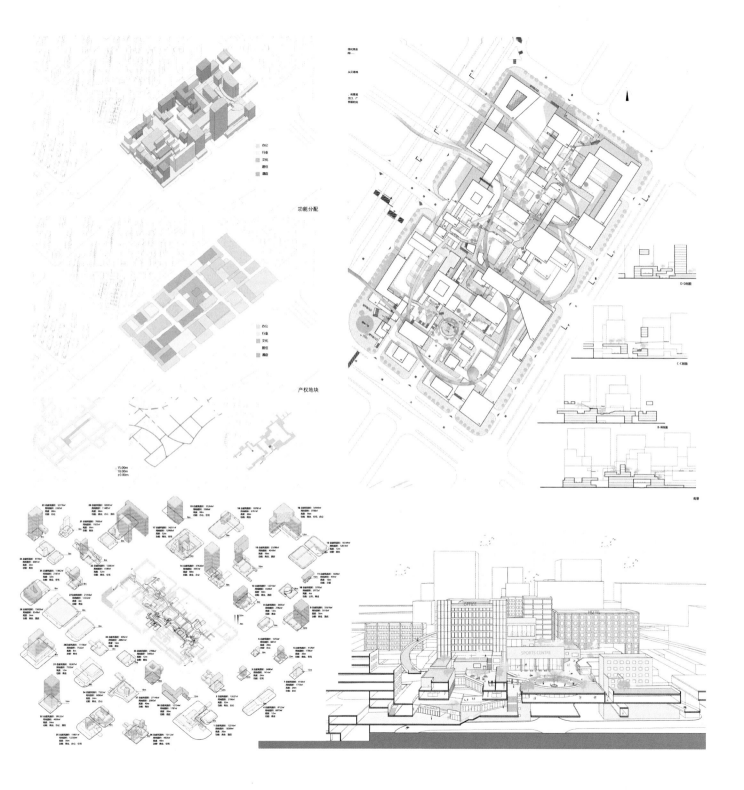

数字化设计与建造
DIGITAL DESIGN AND CONSTRUCTION
吉国华 李清朋

本科毕业设计 UNDERGRADUATE GRADUATION PROJECT

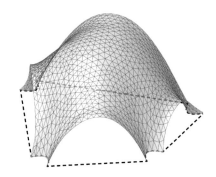

1.教学目标

基于建筑数字化技术，本毕业设计涵盖案例分析、设计研究以及建造实践三个部分，建立基于力学生形的设计方法，解决数字化设计与实际建造的真实问题，完成从形态设计到数字化建造的全过程。整个课程以结构性能为形态设计的出发点，协同思考形式美学与建造逻辑的关系，培养学生在建筑设计阶段主动考虑结构逻辑的能力，在建筑形式创新和结构逻辑之间寻求统一。

2.教学内容

基于力学生形的数字化设计与建造。

3.课题简述

"建筑与结构的关系"是建筑学与建筑设计中最基本、最核心的问题之一。建筑与结构在空间的围合、形体的构筑、形象的塑造等三个方面具有密不可分的关系。从力的感知到受力体系的选择，从结构骨架的支撑到空间形态的实现，从空间形态到建筑作为人类生活空间的容器，基于力学原理的形态设计为建筑空间的设计提供了一个有力的切入点，大大延伸了建筑与结构协同的操作范围，同时也提供了一个完成设计到建造的起点。

本课题以"基于力学生形的数字化设计与建造"为主题，要求学生在学校自选环境中设计一处用地面积4m×4m，遮盖面积为10 m²左右的建筑空间，以满足师生停留、休憩、交流的功能需求。课题通过实物模型制作来不断探索设计问题，用数字化的方法研究和解决问题，最终通过数控加工的方式来实现具有真实细节的构筑物。

4.重点问题

1）基于计算性设计的技术与思维对建筑问题进行解析。
2）结构性能设计的力学原形分析与应用。
3）Grasshopper程序及编程学习，运用各种程序方法和各类库文件。
4）材料研究，充分挖掘并整理与数控建造相关的各类材料。
5）掌握相关模型制作工具（激光雕刻机、CNC、3D打印机、机械臂等）的基本知识和操作要领。

5.成果要求

技术分析图纸与文本、设计文本、实物模型。

6.教学进度

1）第一阶段：案例分析——力学生形（第1~3周）。
第1周：数字化设计原理与思维、力学生形原理与思维讲解；
第2~3周：通过案例的研究学习与模型搭建，提取案例的结构原型，分析其力学逻辑，并建立设计—建造两者之间的互动思维。
该阶段设计工具为：Rhino、Grasshopper与手工模型。
模型制作工具为：激光雕刻机、CNC、3D打印机、机械臂等。
2）第二阶段：设计研究——从原形到设计（4~11周）
第4~5周：面对实际场地，把材料性能、力学逻辑、节点构造等作为设计的出发点，并考虑如场地、功能、空间等典型的建筑设计限定要求，创造出新的数字化形式。
第6~11周：协同思考形式设计与实际建造，考虑方案的可建造性，完成优化并确定方案。
该阶段设计工具为：Rhino、Grasshopper与手工模型。
模型制作工具为：激光雕刻机、CNC、3D打印机、机械臂等。
3）第三阶段：建造实践——从材料到构造（第12~16周）
第12周：选取1~2个节点，试做足尺模型，以对建造进行深入研究与改进。
第13~14周：进行缩尺模型的整体搭建，目的是以此验证设计的整体建造性能。
第15~16周：完成最终展示建筑实物模型以及视频、图纸表达。
该阶段设计工具为：Rhino、Grasshopper。
模型制作工具为：激光雕刻机、CNC、3D打印机、机械臂等。

1.Teaching objective

Based on the digital technology of architecture, this graduation project covers case analysis, design research and construction practice. It intends to establish a design method based on the principle of formation according to mechanics, so as to solve the problems of digital design and actual construction, thus completing the whole process from morphological design to digital construction. The whole course takes structural performance as the starting point of form design, and coordinately considers the relationship between formal aesthetics and construction logic, to cultivate the students' ability to actively consider structural logic at the stage of architectural design, and seek unity between architectural form innovation and structural logic.

2.Teaching contents

Digital design and construction based on formation according to mechanics.

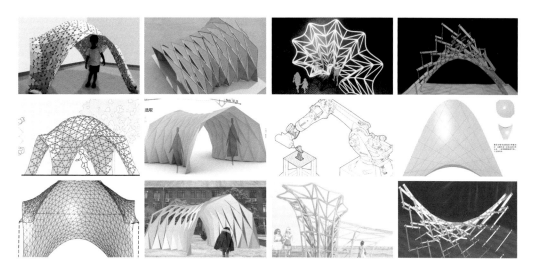

3. Brief introduction

The "relationship between architecture and structure" is one of the most basic and core issues in architecture and architectural design. Architecture and structure are inseparable in aspects of space enclosure, form construction, and image shaping. The form design based on mechanics principle provide a powerful entry point for architectural space design from the perception of force to the selection of the stress system, from the support of structural skeleton to the realization of spatial form, and from spatial form to architecture as the container of living space, which can greatly extend the range of collaborative operation between architecture and structure, and also provide a starting point from design to construction.

Themed in "the digital design and construction based on the formation according to mechanics", this course requires each student designing the architectural space with the site area of 4 m×4 m, and covering area of 10 m^2 in the school, to enable the teachers and students to stay, rest and communicate. It continues to explore design issues through developing physical models, study and solve problems by a digital method, and finally construct the structure with real details by CNC.

4. Major issues

1) Analysis of architectural issues based on the technology and thinking of computational design.
2) Analysis and application of mechanical prototype for structural performance design.
3) Grasshopper program and programming, and the application of various programming methods and library files.
4) Material research, for fully exploring and sorting various materials related to CNC construction.
5) Mastering of basic knowledge and operating essentials related to the modeling tools (laser engraving machine, CNC, 3D printer, and robotic arm).

5. Achievement requirements

Technical analysis drawings and texts, design texts, and physical models.

6. Teaching process

1) The first stage: Case analysis—Formation according to mechanics (week 1~3).

Week 1: Digital design principle and thinking, principle of formation according to mechanics and thinking interpretation;

Week 2~3: Extract the structure prototype, analyze the mechanical logic and establish interactive thinking between design and construction through case study and model building.

The design tools used at this stage include: Rhino, Grasshopper and the manual model.

The modeling tools include: Laser engraving machine, CNC, 3D printer, and mechanical arm.

2) The second stage: Design research—From prototype to design (Week 4~11).

Week 4~5: Based on the actual site, take the material property, mechanical logic, and node structure as the starting points of design, and consider typical architectural design restrictions (site, function, and space), so as to create a new digital form.

Week 6~11: Cooperatively think about the relationship between formal design and actual construction, consider the constructability of the scheme, complete the optimization and determine the scheme.

The design tools used at this stage include: Rhino, Grasshopper and the manual model.

The modeling tools include: Laser engraving machine, CNC, 3D printer, and mechanical arm.

3) The third stage: Construction practice—From material to structure (Week 12~16).

Week 12: Select 1~2 nodes, prepare a full-scale model, for in-depth research and improvement of the construction.

Week 13~14: Carry out overall construction of the scale model, to verify the overall construction performance of the design.

Week 15~16: Finally display the physical model, video and drawing.

The design tools used at this stage include: Rhino, and Grasshopper.

The modeling tools include: Laser engraving machine, CNC, 3D printer, and mechanical arm.

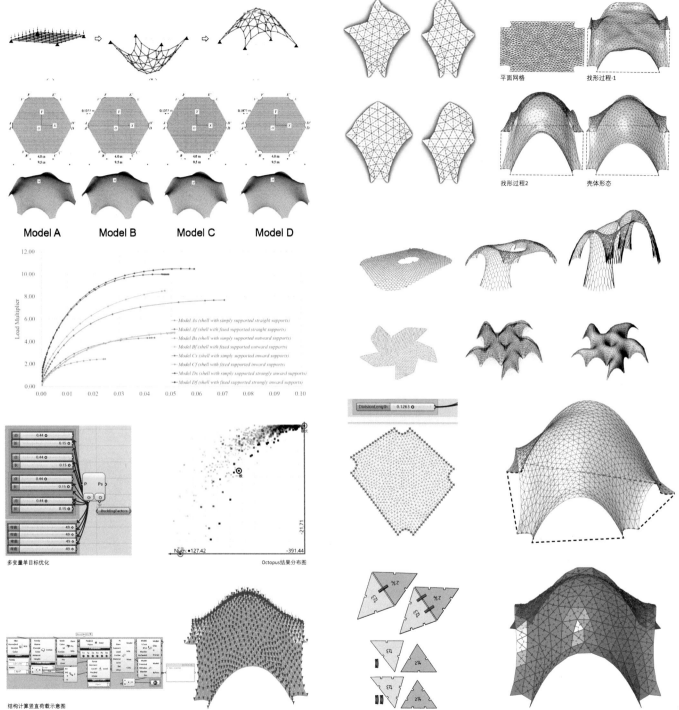

  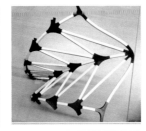  
  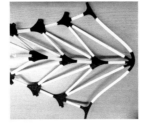  

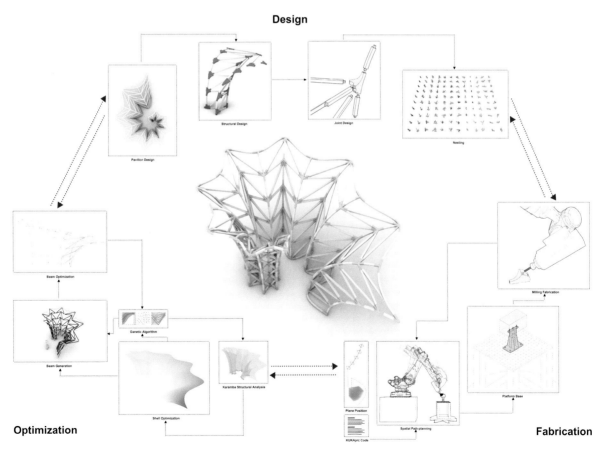

Design

Optimization

Fabrication

空间作为能动者：海上茶路的技术传播与东南亚近代聚落景观的塑造
SPACE AS AGENT: TRANSMISSION OF TEA TECHNOLOGY ALONG THE MARITIME TEA ROAD AND THE SHAPING OF MODERN SETTLEMENT LANDSCAPE IN EAST AND SOUTH ASIA

黄华青

1. 题目简述

空间的"能动性"是建筑学与人类学共同关注的跨学科话题，体现于空间形式的自治性和空间—社会的互动建构，进而成为跨越主体与客体、个体与集体、想象与实体研究的桥梁。本研究从列斐伏尔、布迪厄、白馥兰的现代空间理论脉络出发，探讨空间作为一种兼具物质性/社会性的"技术"传播媒介，如何作为"能动者"而构成并形塑近代聚落的物质和社会景观。

本课题在"一带一路"、聚落文化、近代遗产等视野下，以18世纪中叶至20世纪初兴盛的海上茶叶贸易之路沿线聚落为载体，探讨茶叶技术（包括建筑、机械等物质技术及人力、资本等社会技术）的传播与东亚、南亚等地近现代聚落景观的形成变迁之间的激烈互动。研究重点可包括但不限于：东亚（中国东南沿海、中国台湾、日本静冈）、东南亚（印尼爪哇/苏门答腊、越南林同/太原）、南亚（印度大吉岭/阿萨姆、斯里兰卡康提）。这些国家和地区大多由西方先发工业国家（英国、荷兰、法国、美国等）的殖民和技术输出而经历工业化进程，茶叶生产和贸易在其聚落景观的近代塑造中起到决定性作用；中国作为茶叶生产技术的源头及海上茶路最重要的起点，也随着贸易和技术流动而主动或被动地融入世界体系的重塑之中。

本课题基于史料和田野调查，从中外聚落形态和建筑类型（如贸易城镇、茶村、茶厂等）比较出发，为东亚与南亚近代聚落景观的空间形塑寻求以"人"为基准的线索，探索中国传统聚落和乡土聚落的海外源头，也试图为"一带一路"等多边发展动议提供历史和理论支撑。

2. 课题要求

本课题涉及建筑学、聚落文化、建筑遗产、民族志、移民史、贸易史、技术史等多领域话题，以史料搜集、田野调查、聚落形态及建筑类型研究为手段，成果以毕业论文形式呈现。人数1~2人。

1. Brief introduction

The agency of space, an interdisciplinary topic commonly concerned by architecture and anthropology, is reflected in the autonomy of spatial form and the interactive construction of space-society. It is a bridge spanning the research of subject and object, individual and collective, imagination and entity. In this research, the modern space theories of H. Lefebvre, P. Bourdieu, and F. Bray are used to discuss how does space, the "technical" communication medium with both materiality and sociality, form and shape the material and social landscape of modern settlements as the "agent". Under the background of "the Belt and Road Initiatives", settlement culture, and modern heritage, the dissemination of tea technology (including material technologies of construction and machinery and social technologies of manpower and capital), and the intense interaction with formation and changes of modern settlement landscapes in East Asia and South Asia would be discussed with the settlement formed along the maritime tea road prospered from the middle of the 18th century to the beginning of the 20th century as the carrier. The research focuses on the following contents (including but not limited to): East Asia (Southeast Coast of China, Taiwan, China and Shizuoka, Japan), Southeast Asia (Java/Sumatra, Indonesia, Lam Dong/Thái Nguyên, Vietnam), South Asia (Darjeeling/Assam, India, and Kandy, Sri Lanka), most of which experienced the process of industrialization due to colonization and technological export of Western advanced industrial countries (UK, Holland, France, and the United States); tea production and trade played a decisive role in shaping of settlements in modern times. As the source of tea production technology and the most important starting point of the maritime tea road, China has actively or passively integrated into the remodeling of this world system with the trade and technology spread.

In this course, the settlement patterns and building types (such as trading towns, tea villages, tea factories) would be compared between China and foreign countries based on historical data and field research, to seek the clues based on "human" for space of modern settlements in East and South Asia, and explore the overseas sources of Chinese traditional settlements and rural settlements, thus providing the historical and theoretical support for multilateral development Initiatives of "the Belt and Road Initiatives".

2. Topic requirements

This topic involves architecture, settlement culture, architecture heritage, ethnography, immigration history, trade history, and technology history, and it is researched by means of historical data collection, field investigation, research of settlement pattern and architectural type. The achievements should be presented in the form of graduation thesis. 1~2 students.

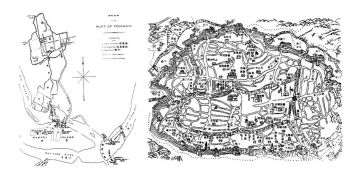

中国茶叶贸易历史大事记

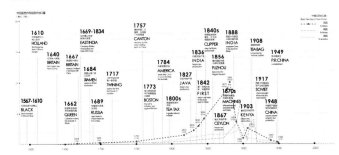

茶叶生产贸易体系流程

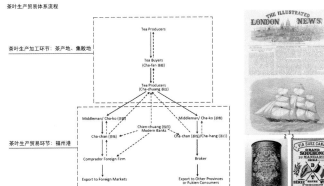

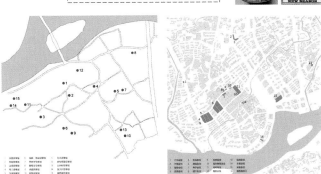

建筑设计研究（一）基本设计 DESIGN STUDIO 1 DESIGN BASICS
住宅兼工作室设计
HOUSE AND STUDIO DESIGN

傅筱

鸟瞰图

1.教学目标

课程从"场地、空间、功能、经济性"等建筑的基本问题出发，通过宅基地住宅设计、训练学生对建筑逻辑性的认知，并让学生理解有品质的设计是以基本问题为基础的。

2.教学过程

1）场地分析

时间：第一周。

内容：分析场地和相关案例研究，并提交分析报告和场地模型（全组一个1：100，用于平时方案讨论，场地范围由学生自行研究确定）。每组分析2~3个案例（包含所选定的结构体系），组与组之间的案例尽量不重复。

2）了解基本建筑设计原理

时间：第二周。

内容：汇报分析报告，并提出初步概念方案，以工作模型或者图纸进行研究，制作PPT汇报。

3）组织空间与行为

时间：第三至六周。

内容：形成解决方案，以工作模型和图纸进行研究，制作PPT汇报。

4）设计研究与表达

时间：第七、八周。

内容：

①完成平、立、剖图纸，比例1：50，平面和剖面要求布置家具、材料填充以及表达人的行为。

②提交场地模型（全组一个1：50，用于答辩）和1：50比例单体工作模型（每组一个）。

③表达空间关系的三维剖透视不少于2个，比例不低于1：50，要求材料填充，必须有家具和人的行为表达。

④有利于表达形体和空间的透视图，可以是渲染图，也可以是模型照片。

⑤其他有利于设计意图表达的图纸。

1. Teaching objective

The course starts from fundamental issues of architecture such as "site, space, function, and economy", aims to train students to cognize architectural logics, and allow them to understand that quality design is based on such fundamental issues.

2. Teaching process

1) Site analysis

Time: Week 1.

Contents: Analyze the site and the related case studies, and submit the analysis report and site model (a model 1 : 100 for each group for scheme discussion; and the site scope should be determined by the students). Each group should analyze 2~3 cases (including the selected structural system), and the cases between groups should not be repeated.

2) Understand the basic architectural design principle

Time: Week 2.

Contents: Submit the analysis report, and put forward a preliminary conceptual scheme; carry out the research based on the working model or drawing, and prepare a PPT report.

3) Organization space and behavior

Time: Week 3~6.

Contents: Form a solution, carry out the research based on the working model or drawing, and prepare a PPT report.

4) Design research and expression

Time: Week 7 & 8.

Contents:

① Complete the plane, elevation and section (scale: 1 : 50), and the plane and section should involve the layout of furniture, filling of materials and expression of human behavior.

② Submit the site model [a model (1 : 50) for each group for reply] and single working model (1 : 50, one for each group).

③ There should be at least 2 three-dimensional sectional perspectives (Scale: No less than 1 : 50) that can express the spatial relationship; the filling of materials, layout of furniture, and expression of human behavior should be involved.

④ There should be a perspective being conducive to expressing the shape and space, which can be a rendering or model photo.

⑤ There should be other drawings being conducive to expressing the design intent.

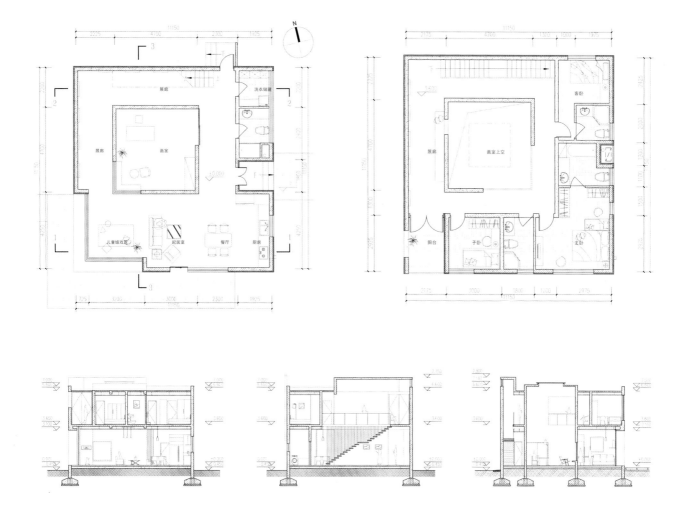

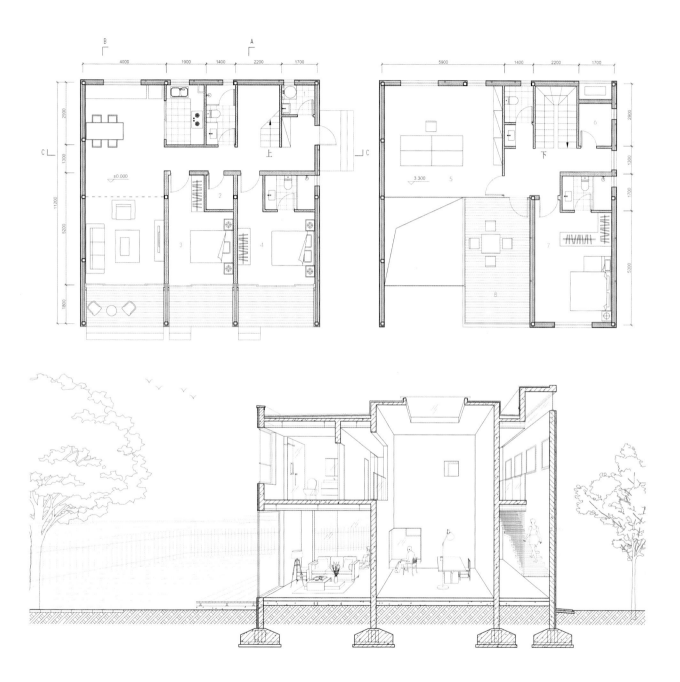

建筑设计研究（二）概念设计 DESIGN STUDIO 2 CONCEPTUAL DESIGN

作为空间实验室的建筑学：文创能力中心设计研究
ARCHITECTURE AS SPATIAL LABORATORY: DESIGN RESEARCH ON COMPETENCE CENTER FOR CULTURAL INNOVATION

鲁安东

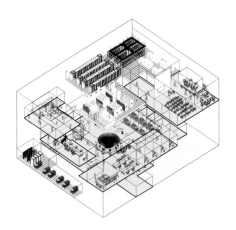

1.教学内容

本课程将探索一种全新的建筑类型，即作为"空间实验室"的建筑，它植根于当代的社会经济、媒体技术和城市空间实践，兼容了物质空间与虚拟空间。本课程将针对文创这一新兴产业开展系统研究，以紫金·创联盟这一产、学、研、资、媒联动机制为切入点，分析当代文创的空间特性和形式创新，并为紫金·创联盟设计一个文创能力中心，使其成为驱动江苏省文化创新的空间实验室。

2.教学目标

本课程将尝试回答下列问题：
1）什么是"空间实验室"？它对建筑有哪些新要求？建筑如何回应这些要求？
2）文创产业如何构成？有哪些主要活动及空间需求？产、学、研、资、媒等要素在其中的各自角色与联动方式。
3）紫金·创联盟如何作为江苏文创产业的支持平台？如何通过空间设计使其成为有效的能力中心？
4）完成文创能力中心改造方案设计。

3.教堂进度

第一周："空间实验室"案例分析及原形研究。
第二周："空间实验室"案例分析及原形研究。
第三周：文创产业研究。
第四周：文创产业研究。
第五周：设计场地及任务书研究。
第六周：方案初步设计。
第七周：方案深化设计。
第八周：评图。

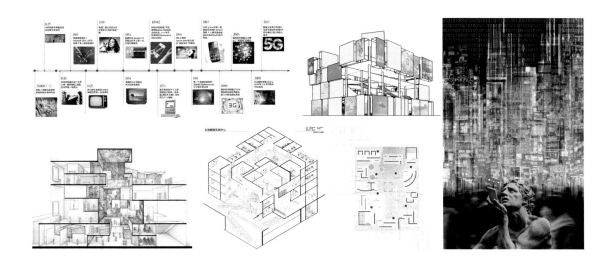

1.Teaching content
This course will explore a new type of architecture, the architecture as "spatial laboratory", which is rooted in contemporary social economy, media technology and urban space practice, and is compatible with physical space and virtual space. It will carry out systematic research based on the emerging industry of the cultural and creative industry, and analyze the spatial characteristics and form innovation of contemporary cultural and creative industry with Purple Creative Union as the entry point; in addition, it will also design a cultural and creative competence center for Purple Creative Union, and make it a spatial laboratory driving cultural innovation in Jiangsu Province.

2.Teaching objective
This course will try to answer the following questions:
1) What is a "spatial laboratory"? What are the new requirements for architecture? How does the architecture respond to these requirements?
2) What is the structure of the cultural and creative industry? What are the main activities and space requirements? What are the roles and linkage methods of industry, university, institute, capital, and media?
3) How does Purple Creative Union serve as a support platform for the cultural and creative industry in Jiangsu Province? How to make it an effective competence center through space design?
4) Complete the design of transformation of competence center for cultural Innovation.

3.Teaching process
Week 1: "Spatial laboratory" case analysis and prototype research.
Week 2: "Spatial laboratory" case analysis and prototype research.
Week 3: Cultural and creative industry research.
Week 4: Cultural and creative industry research.
Week 5: Design site and task book research.
Week 6: Preliminary scheme design.
Week 7: Deepened scheme design.
Week 8: Drawing review.

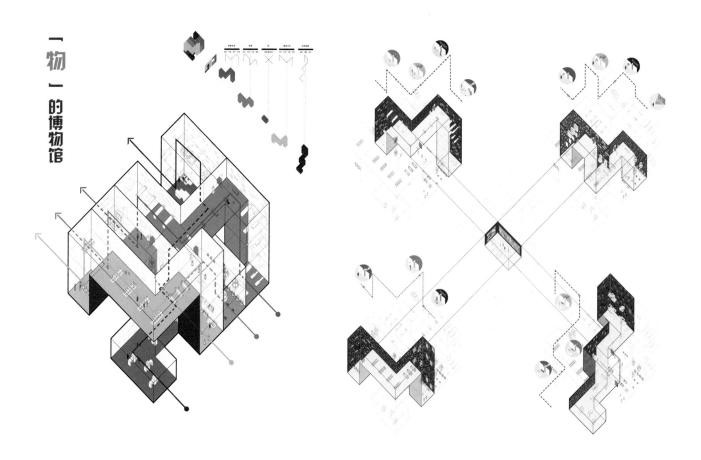

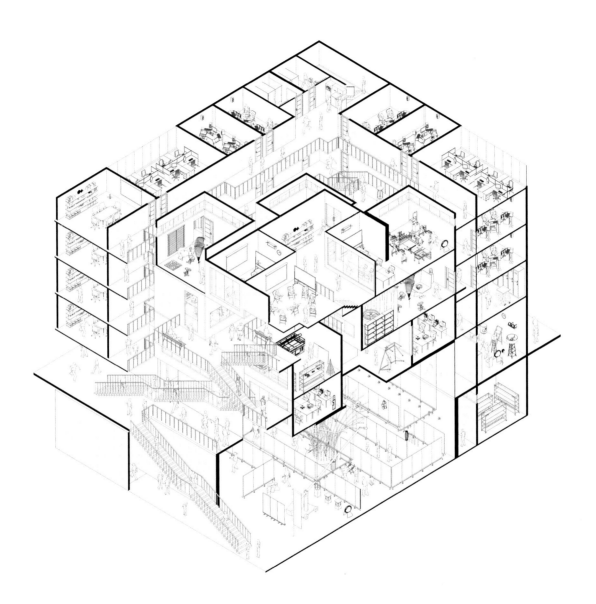

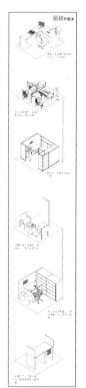
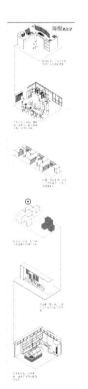

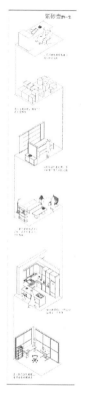

公
共
空
间
中
人
的
活
动

市民进行公共生活，产生聚集　　当权者对王权进行展示，具有仪式性　　现代生活的差异性，形成初步的分散　　现代主义"物体"的城市，导致广场的空荡　　当代信息的发达提升了人对于私密感的要求，呈现出个体分散的状态

公
共
空
间
的
界
面

界面围合度很高，产生集聚效应　　界面成为王权展示的舞台布景，产生了初步室内化的倾向　　界面的高度提升，更具多样性，但横向围合不足　　界面缺乏围合感导致了公共空间仅作为一种交通　　界面呈现全围合的状态，完成了室内化的转变

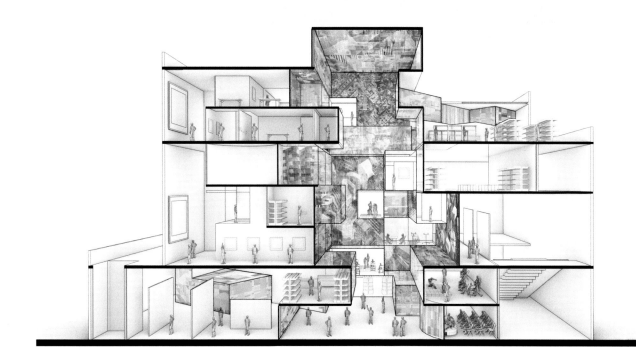

建筑设计研究（三）城市设计 DESIGN STUDIO 3 URBAN DESIGN
国际和平城市空间设计研究
DESIGN RESEARCH ON THE INTERNATIONAL CITY OF PEACE
鲁安东

1.教学内容

本次课程在"南京国际和平城市建设方案"框架下，针对"内容导向的城市设计"进行研究。在城市存量更新的背景下，内容导向的城市设计正变得日益重要，它对于城市的品牌塑造和资源整合、对于城市的文化建设和社区营造都是重要途径。本次课程将探索不同形式的记忆载体如何作为空间设计和场所设计的核心，探索物质空间设计与其他媒介、技术干预的整合可能性。在此基础上，对南京国际和平城市建设提供前瞻性的可行方案。

2.教学进度

研究阶段：
1) 案例研究：对和平的空间诠释（1周）。
对成功结合和平主题与城市建设的国外实例进行系统搜集和分析，形成案例库。
2) 本体研究：XXX作为记忆的载体（1周）。

例如：
类型一：记忆场所。
类型二：无名之物。
类型三：隐形边界。
类型四：个体记忆。
其他自定义类型。
3) 载体研究与空间调查：在调研基础上，绘制和平城市空间资源地图（1~2周）。

设计阶段：
4) 概念性方案设计（1周）。
5) 深化设计（1周）。
6) 制图表现（1~2周）。
7) 国际和平城市设计展。

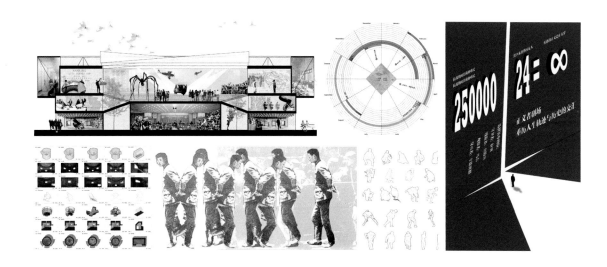

1.Teaching contents
Under the framework of the "Construction Scheme of Nanjing International City of Peace", the research will be carried out on the "content-oriented urban design". The content-oriented urban design has been more and more important under the background of urban inventory renewal, and it is also an important method for brand building, resource integration, urban cultural construction and community building. This course will explore how different forms of memory carriers act as the core of space design and place design, and also explore the possibility of integration of physical space design with other media and technological interventions. On such a basis, it will provide a forward-looking and feasible scheme for the construction of Nanjing international city of peace.

2.Teaching process
Research stages:
1) Case research: Spatial interpretation of peace (1 week).
Systematically collect and analyze foreign examples in successful combination of peace with urban construction, and form a case library.
2) Ontology research: XXX as the carrier of memory (1 week).
For example:
Type 1: Memory place.
Type 2: Indefinable things.
Type 3: Invisible boundary.
Type 4: Individual memory.
Other custom types.
3) Carrier research and space research: Based on investigation, draw a map of spatial resources of the international city of peace (1~2 weeks).
Design stages:
4) Conceptual scheme design (1 week).
5) Deepened design (1 week).
6) Drawing performance (1~2 weeks).
7) Design exhibition of the international city of peace.

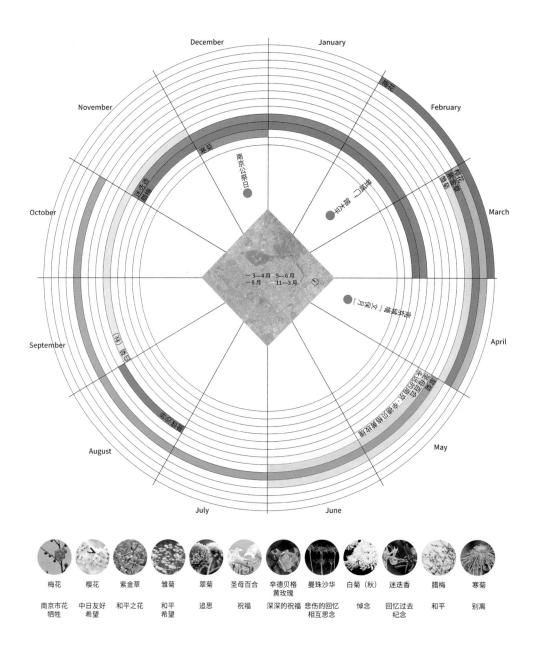

南京城墙现状调研

— 历史
— 和平

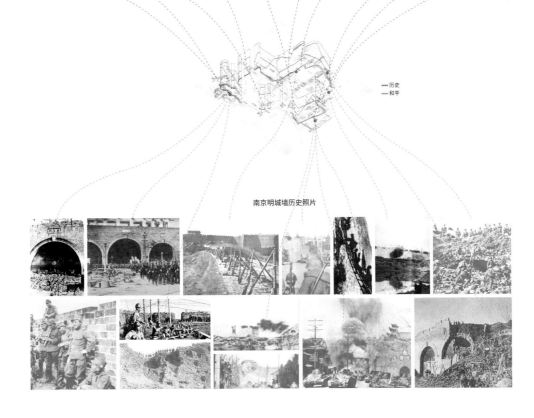

南京明城墙历史照片

建筑设计研究（三）城市设计 DESIGN STUDIO 3 URBAN DESIGN

策划-规划-建筑-景观整合设计
PLOTTING, PLANNING, ARCHITECTURE AND LANDSCAPE INTEGRATED DESIGN TRAINING

周凌

1.教学目标

通过一个具体实际项目的设计训练，掌握策划、规划、建筑、景观等多专业的设计技能与相关知识。具体涉及产业策划、旅游策划、产业规划、环境规划、交通规划等策划与规划方面的知识，涉及建筑设计的基本的结构构造知识、功能组织、形体设计技能。通过课程设计，掌握规划建筑专业各种相关的知识和技能，同时训练各种形式的表达。

2.设计题目

课程将在以下题目中选择一个深化设计。

题目一：南京六合龙袍新城长江社区规划与建筑设计

1）油罐美术馆；
2）农业奥特莱斯；
3）农业会展中心；
4）村民礼堂；
5）农房改造/新农房设计。

题目二：浙江嘉兴东浜头画家村（丰子恺故里）规划改造设计

1）活动中心；
2）展示中心；
3）民宿酒店；
4）理想村乡村公寓。

3.设计内容与成果

1）策划：包含产业、土地、旅游、投资开发模式的策划。
2）规划：包含生态保护、土地利用、交通规划、环境整治、房屋改造、景观治理等内容。
3）建筑设计：包含旧建筑改造、农宅改造、新农宅设计、新配套公建设计等。
4）成果要求：每组8张A1，文字说明不少于2000字，PPT陈述。

4.教学进度

第1周：现场调研，现状分析。
第2周：策划方案与概念规划。
第3周：完成规划。
第4周：建筑概念设计。
第5~6周：建筑设计。
第7~8周：成果表达制作。

1.Teaching objective

Help the students to master design skills and knowledge of plotting, planning, architecture and landscape through design training of a specific project, which specifically involves the knowledge of industrial plotting, tourism plotting, industrial planning, environmental planning, and transportation planning, as well as the basic structural knowledge, functional organization, and shape design skills in architectural design. This course will train various forms of expression in addition to the help of students to master various knowledge and skills of planning and construction through the course design.

2.Design subject

The course will select one of the following subjects for deepened design.

Subject I: Planning and Architectural Design of Changjiang Community, Luhe Longpao New City, Nanjing

1) Tank Museum;
2) Agricultural Outlets;
3) Agricultural Exhibition Center;
4) Villager Auditorium;
5) Rural Housing Renovation/New House Design.

Subject II: Planning and Reconstruction Design of Dongbangtou Painter Village (Hometown of Feng Zikai) in Jiaxing, Zhejiang Province

1) Activity Center;
2) Exhibition Center;
3) Homestay Hotel;
4) Lixiang Village Apartment.

3.Design contents and achievements

1) Plotting: Including the plotting of industry, land, tourism, investment development models.
2) Planning: Including ecological protection, land use, transportation planning, environment renovation, house renovation, and landscape management.
3) Architectural design: Including old building renovation, rural housing renovation, new rural housing design, and design of new supporting public buildings.
4) Achievement requirements: 8 A1 drawings of each group, with text description of no less than 2000 words, and PPT statement.

4. Teaching process

Week 1: Site investigation, and current situation analysis.
Week 2: Plotting scheme and conceptual planning.
Week 3: Completion of planning.
Week 4: Architecture concept design.
Week 5~6: Architectural design.
Week 7~8: Preparation for achievement expression.

总平面图
0 200 500m

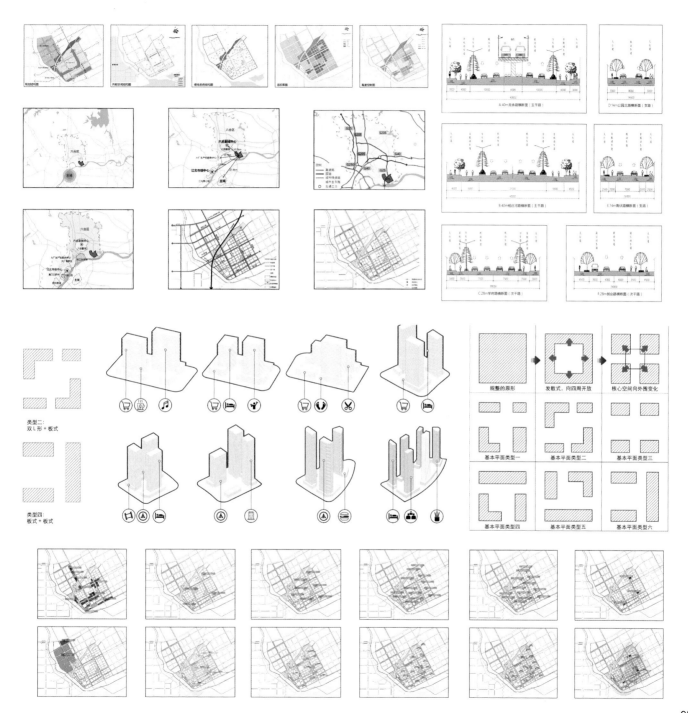

建筑设计研究（三）城市设计 DESIGN STUDIO 3 URBAN DESIGN

内边缘2.0：城市交通廊道综合再利用
INNER EDGE 2.0: COMPREHENSIVE REUSE OF URBAN TRAFFIC CORRIDORS

胡友培

如果说传统城市是一种二维的、经纬交织的水平面，当代城市似乎越来越呈现出一维或线性的特征。线性的要素在当代城市的运转、组构中，扮演着至关重要的角色。内边缘是一种典型的线性要素。它是城市内部各板块之间的缝隙，拥有独特的景观与生态。一种区别于普通城市地貌的临时性、无序性与开放性，使其适合作为各种城市建筑乌托邦的游戏场，进行不切实际、天马行空的设计思想实验。交通廊道及其沿线地带是一种常见而重要的内边缘类型。现代大型交通基础设施盘踞、穿梭在城市板块之间，占据了大量的土地，形成瞩目的边缘性景观。数量巨大的人流、车流、物流在其上不间断地快速流通，城市机体得以正常运转和持续繁荣。它们是当代城市的伟大发明，是根基也是荣耀，是名副其实的基础设施。

1.背景与问题

交通廊道及其沿线地带，往往是工程学的领地。线性交通设施（轻轨、高铁、快速路、高速路、省道、国道等）在其中占据主导地位。基础设施的剖面与线位，被反复推敲优化，形成极端高效的形态，一种工程的美学。

与此同时，由于其位于城市板块的内部边缘，往往被传统设计实践忽视，各种边缘性要素，如厂房、交通附属、村落、林地、耕地等，无序而随机地散落在沿线地带中，形成了一种面目模糊而尺度巨大的城市景观。城市设计与建筑学的专业词汇，如比例、场所、尺度、秩序、界面等，在这里是无效的。作为一种"非地方、非场所"，交通廊道及其沿线地带，似乎从来没有也难以成为城市设计与建筑学的对象。

另一方面，随着中国城市建设进入存量发展的阶段，大城市的土地资源日益稀少而珍贵。交通廊道及其沿线用地尽管利用难度高，但因其不可忽视的存量，而具有极大的再利用与综合开发价值。

如何以一种设计或建筑学的方式来处理这种独特的城市景观和基础设施？建筑学如何参与到当代城市的基础建构工作中？这是课程需要探索并试图回答的中心问题。与此同时，保持创造的轻松，保持现实的批判，以一种既游戏又严肃的姿态开展设计与研究，是课程的基调。将边缘性的物质要素纳入设计学科的范畴，既是一种跨越学科边界的逃逸，也是一种对当代建筑学新命题的审视。

2.任务与要求

场地研究：课程在南京都市区内选择四处典型的交通廊道地带，作为设计研究的场地。研究场地在城市化进程中的变迁，阐明基础设施在其中的主导地位，并分析呈现各种伴生性的城市问题。

场地愿景：从城市发展的全局视角，研判场地的价值与潜力，为场地建构新的身份，提出新的愿景。制订场地的功能计划与策略性的项目介入，探索场地再利用的可能与路径。

原形设计：课程将以原形设计为主要的设计工具。通过对建筑、基础设施、景观以及它们的混杂，开展原形设计，赋予场地某种适宜的、具有想象力的空间形态。

If a traditional city is deemed as a two-dimensional horizontal plane with intertwined latitude and longitude, a modern city may show more one-dimensional or linear characteristics. Linear elements play an important role in operation and organization of a modern city. Inner edge (I.E.) is a typical linear element, which is the gap between various plates of a city, with unique landscape and ecology. It is a temporary, disordered and open part different from ordinary urban areas, making it a playing field for unrealistic and unconstrained design experiments of urban buildings. Urban transportation corridors and the surrounding areas are common and important inner edges. Large-scale modern transportation facilities are entrenched and shuttled between different urban blocks, and occupied a large amount of land, forming eye-catching edge landscape. A large amount of people, vehicles, and logistics may be transferred uninterruptedly and quickly, enabling the city to realize normal operation and continuous prosperity. They are great inventions for modern cities, and also the foundation and glory, as well as veritable infrastructure.

1.Background and problems

Each urban transportation corridor, as well as its surrounding area, is always a domain of engineering. Linear transportation facilities (light rail, high-speed rail, highway, expressway, provincial road, and national road) occupy a dominant position. The section and route position of each infrastructure have been repeatedly refined and optimized, so as to form an extremely efficient form, namely, engineering aesthetics.

At the same time, due to the location at the inner edge of an urban block, it is often ignored by traditional design practices. Various edge elements, such as plants, traffic attachments, villages, woods and cultivated land, scatter randomly along the route, forming a fuzzy and huge-scale urban landscape. The terms of urban design and architecture, such as proportion, place, scale, order, and interface, are invalid. The urban transportation corridors and their surrounding areas, as the "nowhere" and "none place", have never been and would not be the subject of urban design and architecture.

On the other hand, with the stock development of urban construction in China, land resources in large cities have been increasingly scarce and precious. The urban transportation corridors and their surrounding areas have great value of reuse and comprehensive development because of the considerable stock, although there is a high difficulty in utilization.

The course should explore and answer two central questions: Which design or architectural method should be used to deal with the unique urban landscape and infrastructure? How does architecture engage in the basic construction of a modern city? At the same time, the course tries to realize relaxed creation, keep realistic criticism, and carry out design and research with a playful and serious attitude. It is an escape across the boundary of disciplines and also the examination of new propositions of contemporary architecture to include edge elements into the category of design.

2.Tasks and requirements

Site study: Select four typical traffic corridors in Nanjing metropolitan area as the site for design and research. Study the changes of the site in the process of urbanization, clarify the dominant position of infrastructure in the process of urbanization, and analyze various accompanying urban problems.

Site vision: Study and judge the value and potential of the site from the overall perspective of urban development, construct a new identity for the site, put forward a new vision, formulate the functional plan of the site and strategic project intervention, and explore the possibility and path of site reuse.

Prototype design: Take prototype design as the main design tool, and give the site a suitable and imaginative space form through prototype design of architecture, infrastructure, landscape and their mixture.

场景呈现 场地诠释

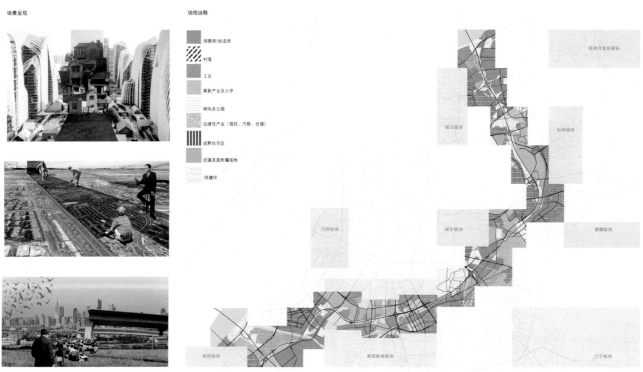

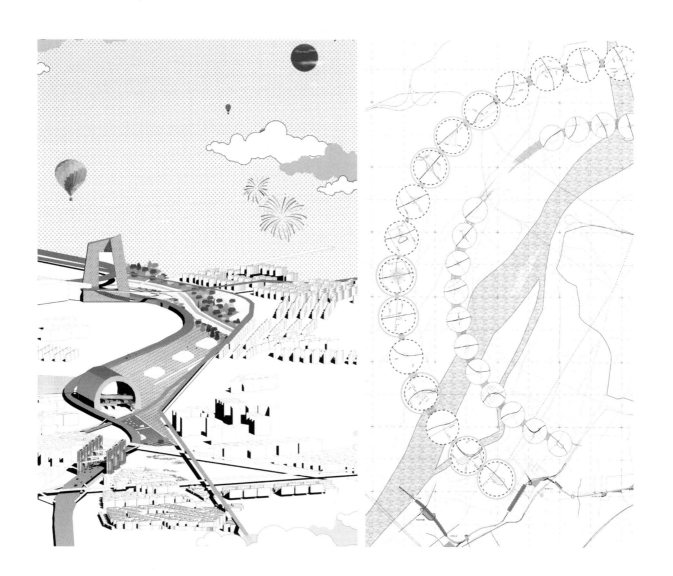

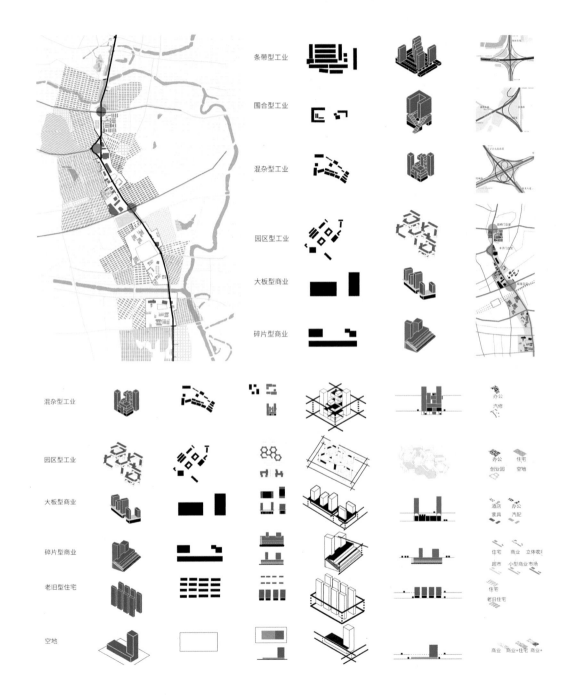

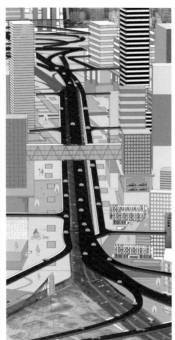

研究生国际教学交流计划 THE INTERNATIONAL POSTGRADUATE TEACHING PROGROM

创新项目：南京周边空间环境再生
INNOVATION PROJECTS: REGENERATIVE ENVIRONMENTS FOR PERIPHERAL NANJING

凯瑞·希瑞斯

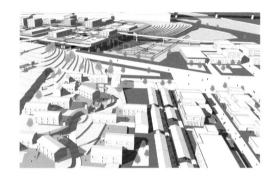

1.教学目标

在本课程中，我们将考虑大多数被私人独栋建筑、功能单一的交通走廊和风景优美的园林区所占用的南京部分周边空间的改造方式。我们计划对改造后周边空间新的用途进行考察，有望在城市扩容过程中利用这些空间来打造关注社会和环境的公民服务机构。为实现这一目标，我们将探索如何创新地将建筑的空间制造能力、基础设施的服务分级和地形景观美化的组合能力联系起来，并将它们整合到一个独特的组合中。通过借鉴这些独特空间生产实践方面的知识，城市设计的作用不再是单单为广大消费者提供标准化空间产品。通过融合不同的设计能力和对建筑设计的敏感性，城市设计可以成为一种从尚未开发的地方和人的潜力中汲取经验的艺术，一个从集体空间中获得最公平表现的技能网络，一个致力于为我们的时代创造适当的表现性聚居地的共享平台。为了在城市设计中实现这样的转变，我们必须考虑未开发规划混合体的潜力，它不仅为生活、劳动和休闲提供了高品质场所，而且还能实现自有资源的再生，从而给城市及其人民创造持久的价值。

本课程将以前瞻性的设计研究模式为指导，并提出这样一个问题：假如这样会如何？在一个快速变化的时代，我们迫切需要新奇的想法和冒险的思维方式——以彻底地改变地球上的居住方式。

2.教学内容

1）介绍"创新项目"和"跨行业订单"的概念，并结合研讨会课程中介绍的理论材料，提供这些概念的论述背景；

2）介绍各种城市干预尺度、场地分析模式、场景建模和时间分段的技术，以及社会和空间领域的概念；

3）介绍绘图、图表、场景建模和模型制作的多种技术，这些技术均为空间表达和设计思想交流的创新手段；

4）介绍与其个人项目提案相关的建筑、基础设施和景观设计方面的参考资料，并向学生提供选定参考项目的概念和实用框架；

5）介绍一系列在专业公共论坛上展示城市研究成果的方法，学生将熟练掌握批判性设计思维，以专业的形式管理其工作，并以英语进行公开展示。

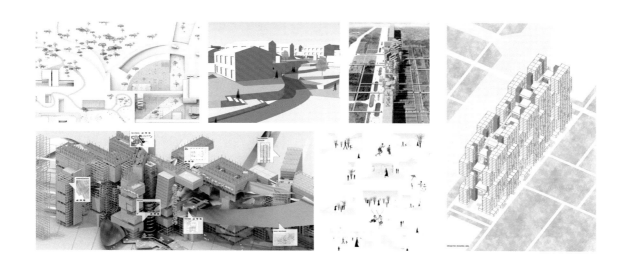

1.Teaching objective

In this course, we will consider how to reclaim a part of Nanjing's periphery, much of which is already lost to privatized, stand-alone buildings, to mono-functional transport corridors, to the nostalgic reserve of scenic landscape areas. We aim to test new uses for the reclaimed peripheral space that could make it perform as a socially and environmentally mindful civic service agent for the expanding metropolitan region. To achieve this goal, we will explore how to innovatively link the spacemaking capacities of architecture, the staging of services by infrastructure, and the compositional abilities of landscaping the terrainsof the earth, and bring them together in a combinatory heterotype . By drawing on the knowhow of these distinct practices of spatial production, urban design could become more than just the delivery of standardized spatial products for a mass audience of consumers. By fusing different competences and sensitivities, urban design could become an art of learning from untapped local potential of places and people, a web of adaptive skills attuned to getting the most equitable performance from collective spaces, and a shared platform dedicated to creating suitably performative habitats for our time. To accomplish such transformations in urban design, we must consider the unexplored potential of programmatic hybrids that not only provide qualitative places for living, labor, and leisure, but also regenerate their own resources and thus give something of lasting value back to the city and its people.

The course will be guided by a prospective mode of design research, asking the question: What if? In an age of accelerating change, we urgently need novel ideas and risky ways of thinking–making that question destructive ways of inhabiting the planet.

2.Teaching contents

1) Students will be introduced to the notion of "heterotypes" as well as "conglomerate orders" and will be given a discursive background of these concepts in conjunction with the theoretical material introduced in the seminar course;

2) Students will be introduced to various urban scales of intervention, modes of site analysis, techniques for scenario modeling and time phasing, as well as the notion of social and spatial territoriality;

3) Students will be introduced to multiple techniques of drawing, diagramming, scenario modeling, and model making as innovative means of spatial expression and communication of design ideas;

4) Students will be introduced to references from architecture, infrastructure, and landscape design relevant to their individual project propositions and will be provided with the conceptual and pragmatic framework for the selected reference projects;

5) Students will be introduced to a range of techniques for professionally presenting urban research findings in a public forum and will be skilled in critical design thinking, curating their work in a professional format, and making public presentations in English.

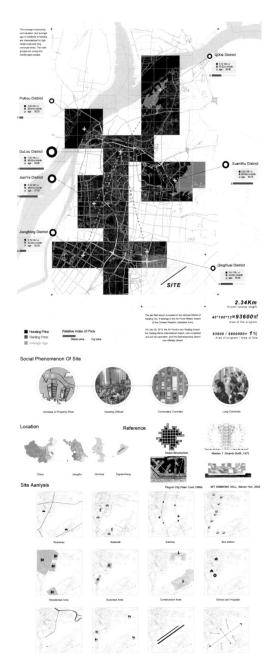

yoUngTOPIA

Youngtopia takes the customized residential unit as the central theme. On the former site of Dajiaochang airport in Qinhuai District, a huge structure along the original airport runway is built. This super building provides a residential framework, and residents can choose to fill in their own residential unit in their favorite location. This is a temporary and dynamic youth community, which is a paradise for youth in this modern and mechanical city.

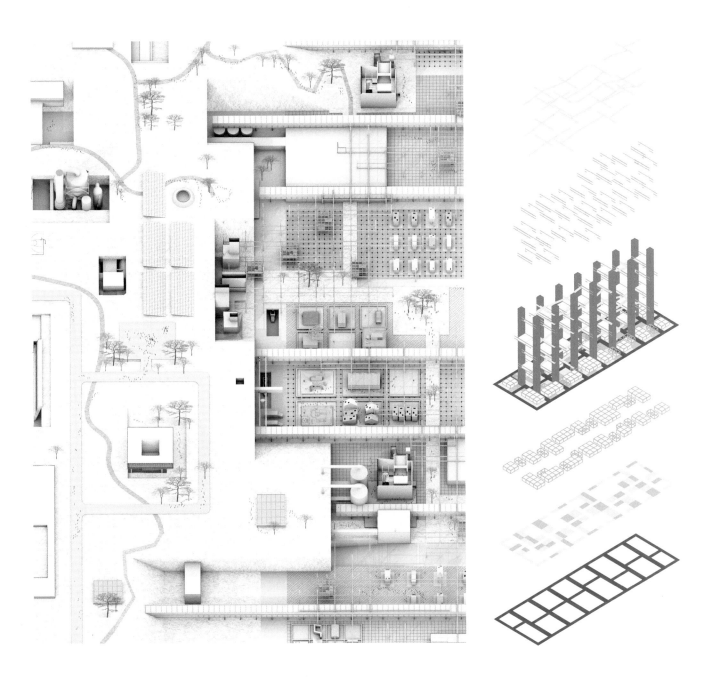

研究生国际教学交流计划 THE INDEROATIONAL POSTGRADUATE TEACHING PROGRAM

让自然融入我们的建筑之中
ALLOWING NATURE TO BE A PART OF OUR ARCHITECTURE

伊斯梅尔·多明格兹

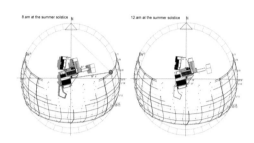

本课程目的是设计出顾及环境条件以及城市、社会和景观的当代建筑。为此，本课程通过四个基础来开发，即案例研究、环境工具、设计方法和策略以及规划行动。

提供"将环境因素融入建筑中的不同方式"的列表，每个学生团队从中选择一个案例研究。将通过双重过滤、符合建筑和可持续的方法对案例进行整体分析。

该分析包括使用计算机模拟工具以及天气条件、方向、类型、内部配置和施工效率。

从该分析中，学生将通过归纳学习不同的策略和设计方法来面对他们的项目，一方面考虑当地和全球的文化价值，另一方面考虑环境质量和高效率表现。

讲座课程将采用每周讲座的方式，主题范围的重点是：为开发高质量和高性能的建筑和城市设计，项目考虑自然流动和环境的重要性和影响。

为此，将介绍多个案例研究，通过从哲学和概念到具体的详细选择来阐述处理环境的不同战略（城市／领土背景、景观、天气条件、社会和文化现状）。本课程还将展示如何分析一个架构件，使其适应可用的信息级别，并提供工具来深入了解它。

This course aims to design contemporary architecture taking into account the environmental conditions as well as urban, social and landscape aspects. For that purpose, the course is developed through four bases: case studies, environmental tools, designing methods and strategies, and projecting action.
Each student team will choose a case study among a list representative of "different ways to integrate environmental aspects into the architecture". They will be analyzed integrally with a double filter, architectural and sustainable approach.
The analysis includes the use of computer simulation tools as well as weather conditions, orientation, typology, internal configuration and construction efficiency.
From this analysis, students will learn by induction different strategies and design methods for facing their projects considering in one hand local and global cultural values, and in the other environmental quality and high efficiency performance.
The lecture course will introduce along weekly lectures, a scope of themes pointing the importance and consequences of permitting natural fluxes and environment to be an integral part of the project, for making high quality and performance architecture and urban design.
For that purpose, several case studies will be presented showing different strategies to deal with the environment (urban/territorial context, landscape, weather conditions, social and cultural realities), from philosophical and conceptual to specific detailed choices. The course will also show how to analyze an architectural piece adapting it to the level of information available and providing tools to go in depth in the knowledge about it.

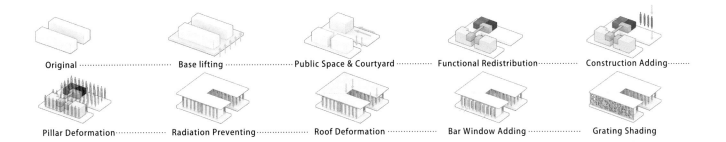

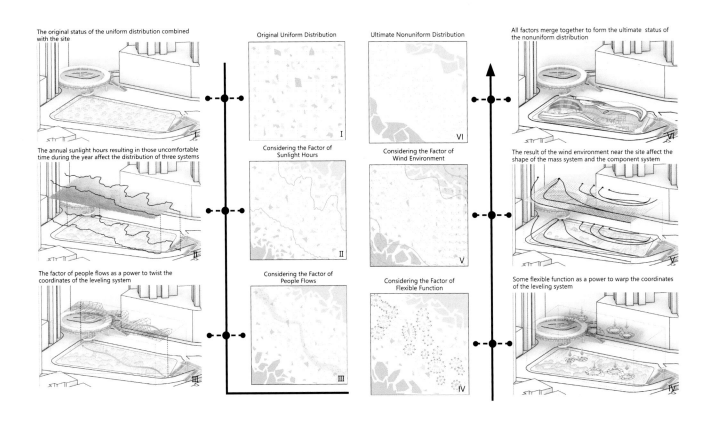

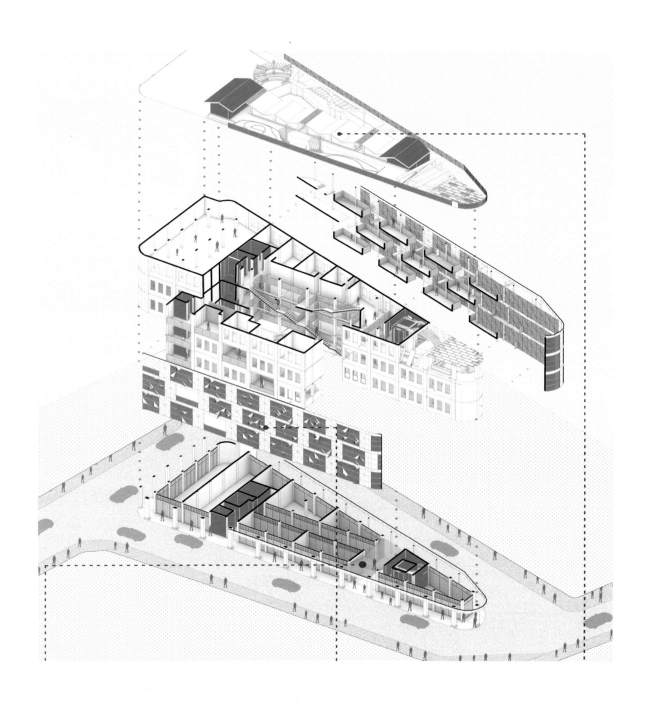

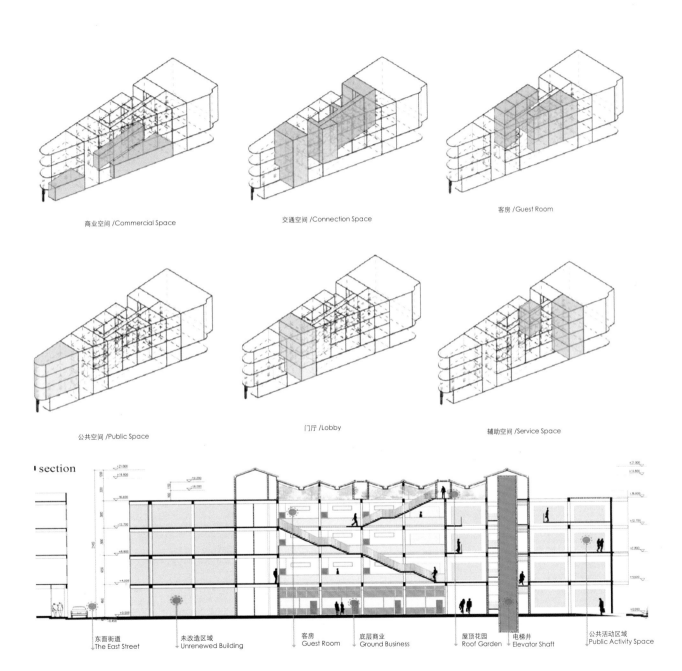

研究生设计工作坊 GRADUATE DESIGN WORKSHOP

古卫城墙的未来可能
THE ANCIENT WALLED CITY AND ITS FUTURE POSSIBILITIES
刘宇扬

在过去的十年间,由于泉州老城的申遗计划,地方政府亦开始在地文化的层面重视历史保护、旅游、华侨返乡和社区营造等。自2014年起,地方政府做过多轮总体规划。而由于对历史文化较为粗浅的理解,在几年前古城的东南侧重建了一段城墙和一座城门。但当地政府也很快地意识到这是花钱建了一个毫无历史价值和意义的假古董。自去年开始,新的策划、规划和设计团队的介入,为当地带来了新的理念和方向,永宁古卫城的保护与发展有了新的契机。

本届工作坊的主题"华侨文化综合体",是一个关于侨乡文化、城墙记忆、文创空间的实验性设计干预。这个设计干预可以兼具改造和新建的建筑、景观和艺术装置的性质,它可以涵盖商业、展示、居住、工作等功能,它需要考虑历史与创新的冲突与融合,它需要处理好动线的贯通和空间的再利用。它需要回应的核心问题是:

如果原始城墙是为历史记忆,现存建筑是为城市记忆,那未来建造是为什么?

意大利建筑师奥尔多·罗西关于城市历史、碎片和记忆的出版论述和实践作品,影响了1970—1980年代的现当代欧洲建筑师群体。参与的研究生同学们可借此工作坊契机,通过课前阅读和课间讨论,结合永宁的城市历史及设计课题,对罗西的理论和实践形成初步认识并获得新的启发。

在为期两周的工作坊中,同学们将通过理论研讨的过程取得一系列的概念关键词,并将概念关键词转化为图像与空间操作。根据工作坊惯例和同学需求,合理安排线上讨论及评图。同学们可结合文字、工作草图、模型、各类表现图纸以及其他辅助视觉表现手段如虚拟现实、电影等,形成一个最终的概念设计干预提案。终期汇报提交一篇大约1000字的文字论述,一份不少于20页的PPT文件(中英文)和一个不超过3分钟的短视频。工作坊旨在从理论/方法论作为工作切入点,从研究到设计建立一套合理的工作方法,重点在于过程中的设计实验和方法探讨,为同学们在日后的学习与实践提供新的思考方向。

In the past ten years, due to the heritage application plan of the Old Town of Quanzhou, the local government has attached great importance to historical preservation, tourism, return of overseas Chinese, and community building. Since 2014, the local government has proposed several rounds of overall planning. Due to a relatively shallow understanding of historic culture, the local government rebuilt a section of walls and a gate on the southeast of the ancient acropolis years ago, but soon, they realized that it was the fake construction without any historical value or significance. Since last year, with the intervention of the new plotting, planning and designing teams, new ideas and directions have been brought, and new opportunities have been created for the protection and development of Yongning Ancient Acropolis. The workshop themed in "Overseas Chinese Cultural Complex" is an experimental design intervention involving culture of overseas Chinese hometown, memory of walls, and cultural and creative space. The intervention can combine the properties of the renovated and newly-built architecture, landscapes and art installations, and it can cover the business, exhibition, residence, and working functions. The conflict and integration between history and innovation should be considered, and the connection of flow and reuse of space should be dealt with. The core question to be responded is:
If the original wall is remained for historical memory and the existing architecture for city memory, then what is the purpose of future construction?
The Italian architect Aldo Rossi published several works about urban history, fragments, and memory, which affected a whole generation of modern-contemporary European architects after 1970s and 1980s. All participating graduate students can take the opportunity to form a preliminary understanding of Rossi's theory and practice and obtain new inspiration through pre-class reading and inter-class discussion in combination with the history of Yongning and the design topic.
In the two-week workshop, students will obtain a series of concept keywords through the process of theoretical discussion, and transform the concept keywords into image and spatial operation. According to the working method and students' requirements, online discussion and drawing evaluation will be arranged reasonably. Students can form a final conceptual design intervention pattern by combining words, working sketches, models, various performance drawings and other auxiliary visual expression means, such as virtual reality and film. In the final report, a text discussion of about 1000 words, a PPT document of no less than 20 pages (in Chinese and English) and a short video of no more than three minutes shall be submitted in the final report. The workshop aims to establish a set of reasonable working methods from theory / methodology through research to design, focusing on the design experiment and method discussion in the process, so as to provide new thinking direction for students' study and practice in the future.

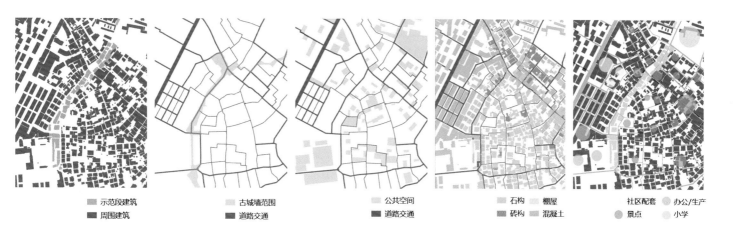

| 示范段建筑 | 古城墙范围 | 公共空间 | 石构　棚屋 | 社区配套　办公/生产 |
| 周围建筑 | 道路交通 | 道路交通 | 砖构　混凝土 | 景点　小学 |

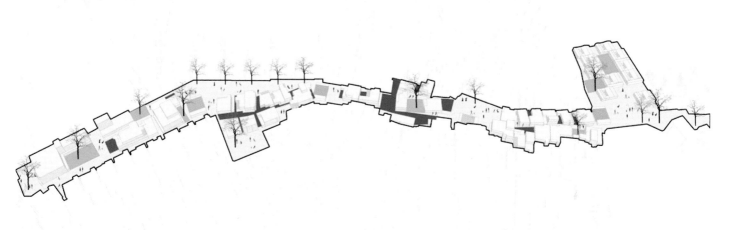

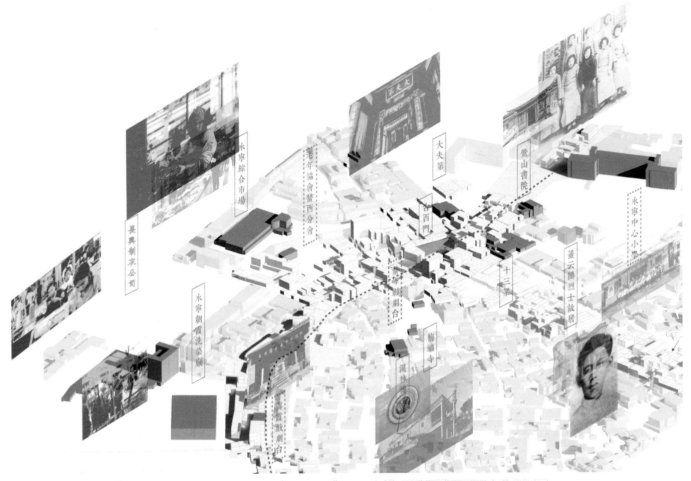

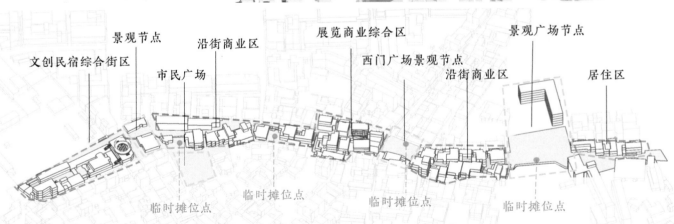

研究生设计工作坊 GRADUATE DESIGN WORKSHOP

力透纸"贝":日常材料的催化利用
FORCE PENETRATING PAPER: CATALYTIC UTILIZATION OF DAILY MATERIALS

朱竞翔

课程第一阶段,探究纸张和书本的性能特征,选择合理的连接方式与几何空间分布形式,进而改善材料的多重性能,实现材料的催化利用。第二阶段,学生沿着教学讨论中阐明的方向继续深入设计,并延伸到相关产品设计,结合特定的人体部位进行可行性的操作。第三阶段,针对特定的产品设计,从材料选择、连接方式和性能提升等方面进行深化、完善设计。第四阶段,通过实验数据、照片、图纸和书面描述,记录工作过程与实践结果。

在理论讲座中,朱竞翔教授首先介绍了罗伯特·马拉尔典型的桥梁工程案例,及其对于拱形管状结构的探索与实践,以此帮助同学们加深对结构性能的理解。其次介绍了其向建筑领域进行的扩展,在薄壳形态的揭示、无梁楼盖的发明、利用平面静力图去理解桁架屋顶结构三个方向进行了探索。香港中文大学的博士生翟玉琨通过建筑师的作品介绍,纸筒、纸板等材料以及插接、折叠等连接方式进行解读,并对比"1958年世博会芬兰馆"与"青海拉吾尕小学"两个项目,体现结构、建造、表皮和空间等多方面的统一的设计思想。

At the first stage of the course, the performance characteristics of paper and books are explored, and reasonable connection method and geometric spatial distribution are selected, to improve the multiple properties of materials and realize their catalytic utilization. At the second stage, all students deepen the design along the direction determined in teaching discussion, extend it to the design of related products, and perform feasible operations in combination with specific body parts. At the third stage, all students deepen and improved the design of specific products in terms of material selection, connection method and performance improvement. At the fourth stage, all students record the work process and practical results based on experimental data, photos, drawings and written descriptions.

In the theoretical lecture, Professor Zhu Jingxiang firstly introduced the typical bridge designed by Robert Maillart in terms of the exploration and practice of the arched tubular structure, aiming to help them to deepen the understanding of structural performance. He secondly introduced the expansion to the architectural field, and made an exploration in the reveal of thin shell shape, the invention of flat slab, and the understanding of roof truss based on plane static. Zhai Yukun, doctoral student of Chinese University of Hong Kong, interpreted the materials such as paper tubes and cardboard and connection methods such as insertion and folding based on introduction of the architect's works. In addition, he compared "Finland Pavilion at World Expo 1958" and "Qinghai Lawuga Primary School", and reflected the unified design integrating structure, construction, surface and space.

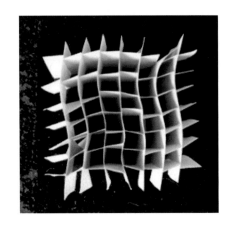 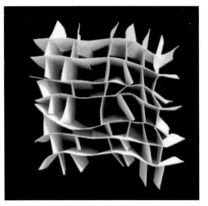 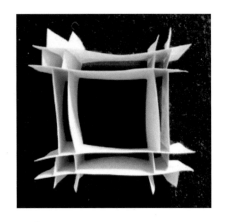

	平面形态	切割与插接			承重能力
四边形构型					约 1.3 kg
三角形构型					约 1 kg
米字形构型					约 1.9 kg

废墟的可能性
THE POSSIBILITY OF RUINS

研究生设计工作坊 GRADUATE DESIGN WORKSHOP

张宇星

1. 教学内容

在今天，废墟越来越成为被热烈讨论的话题。一方面是由于互联网虚拟空间正在成为真实物理空间的全面替代者。互联网空间的内在废墟性，使得我们的当代城市日趋废墟化。从商业街到购物中心、从中央商务区到游乐园、从零售店到街头摊贩，无不正在抵抗着虚拟世界的强烈冲击。另一方面，废墟美学也成为未来城市的一种可能性场景，因为废墟包含了时间性和自然性，也包含了去结构化与去中心化的内在力量，它们都与未来数字世界的建构逻辑直接关联。所有这些都注定了，与废墟相关的理论和实践将成为下一代建筑学需要重点关注的学术命题。

本次工作坊分为三个部分：理论教学、设计研讨、答辩评图。

课程以深圳宝安区蚝乡湖旧电厂为案例，要求学生初步接触废墟的基本概念，梳理相关案例，学会把废墟的思想理念纳入旧建筑改造的设计过程中。通过废墟意向设计·拓展设计思维，为今后的旧建筑改造设计实践打下基础。更进一步，延展思考有关"废墟建筑学"的理论，特别是对废墟所包含的建筑学本体属性。这样的深度思考将有助于学生拓宽视野，未来可以进入更加广阔的学术领域。

2. 教学进程

2020.7.1：讲解废墟与建筑的基础理论、介绍设计课题；
2020.7.4：每组汇报相关案例、蚝乡湖旧电厂改造想法，确定设计方向；
2020.7.7：汇报研讨一轮草图方案；
2020.7.10：汇报研讨一轮深化设计方案；
2020.7.15：汇报研讨二轮深化设计方案，确定小论文框架、出图逻辑；
2020.7.18：14：00，邀请嘉宾、答辩评图。

本工作坊一共18位学生，每组3人，分成6组。最后提交设计图纸PPT与小论文一篇。

1. Teaching contents

Nowadays, ruins are becoming a hot topic. On the one hand, Internet virtual space is becoming a comprehensive substitute for real physical space. The inherent ruins of the Internet space make our contemporary city increasingly ruins, from commercial street to shopping mall, from CBD to amusement park, from retail stores to street vendors, all are resisting the strong impact of the virtual world. On the other hand, the aesthetics of ruins has become a possible scene of the future city, because the ruins contain the timeliness and naturalness, as well as the internal strength of unstructured and decentralized, which are directly related to the construction logic of the future digital world. All these are doomed to the theory and practice related to ruins, which will become the academic proposition that the next generation of architecture should focus on.

This workshop is divided into three parts: Theoretical teaching design discussion, defense plan.

The course takes the old Haoxianghu Power Plant of Bao'an District, Shenzhen as the research case; the students are asked to preliminarily understand the basic concept of ruins, sort out the related cases, and learn to incorporate the thinking of ruins into design of renovation of old buildings. They are expected to expand the design thinking based on intended design of ruins, so as to lay a foundation for renovation design of old buildings. Furthermore, the students are expected to extend the theory of "Ruins Architecture", especially the properties of buildings contained in ruins. The in-depth thinking will help the students to broaden the vision, and enable them to step into a broader academic field in the future.

2. Teaching process

July 1, 2020: Teach the basic theories of ruins and architecture, and introduce the design topics;
July 4, 2020: Report the related cases by groups, introduce the ideas for renovation of old Haoxianghu Power Plant, and determine the design direction;
July 7, 2020: Report the draft of the first round of discussion;
July 10, 2020: Report the deepened design scheme of the first round of discussion;
July 15, 2020: Report the deepened design scheme of the second round of discussion, determine the essay framework, and drawing logic;
July 18, 2020: 14: 00, invite the guests for reply and drawing review.

This workshop consists of 18 students, who are divided into 6 groups (3 students in each group). Finally, each group should submit a design drawing (PPT) and an essay.

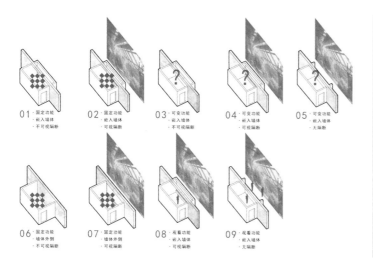

未完待续…

  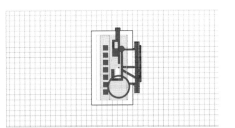
 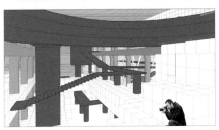 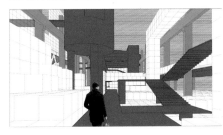

—Vision Of Tomorrow—

废墟化与空间结构显现
——废墟生成机制研究

墙里墙外，
墙外，
汽车飞驰，
须臾败落，

墙里，
水平如镜，
树荫枯荣，
花木墙垣，
洞存生趣，
我与时间共处。

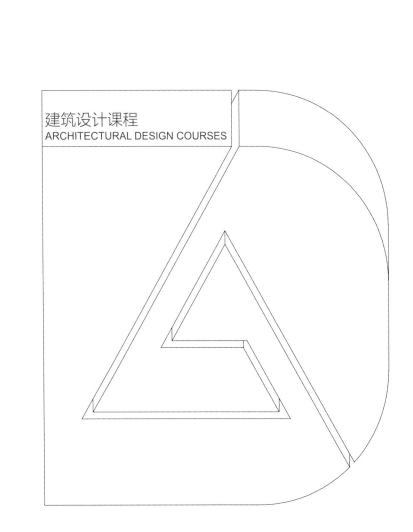

建筑设计课程
ARCHITECTURAL DESIGN COURSES

本科一年级
设计基础
·鲁安东　唐莲　尹航　梁宇舒
课程类型：必修
学时学分：64学时 / 2学分

Undergraduate Program 1st Year
BASIC DESIGN · LU Andong, TANG Lian, YIN Hang, LIANG Yushu
Type: Required Course
Study Period and Credits: 64 hours/2 credits

教学内容
本教案基于四条主题线索和三个能力培养阶段，设计了12个时长五周的教学模块。学生可以根据自己的兴趣和需求自由选修不同模块，量身塑造自己的设计思维和设计能力。通过模块化教学，本教案发挥了通识教育下自主学习的优势，开展理性、全面的思维训练，突出系统、多元的能力培养。通过将设计基础作为创意工科的"元"学科，既为学生进一步的专业学习打下扎实基础，也培养了学生未来跨学科创新的必要素质。

Teaching content
In this teaching plan, 12 five-week teaching modules are designed based on the four thematic clues and three competence training stages. The students can freely select different modules according to their own interests and needs, to tailor their design thinking and design competence. Through modular teaching, this teaching plan can make use of the advantages of autonomous learning under liberal education, thus carrying out rational and comprehensive thinking training, and highlighting systematic and multivariate competence training. Basic design, the "meta" subject of creative engineering, can lay a solid foundation for professional learning by the students, and cultivate the necessary competences for interdisciplinary innovation in the future.

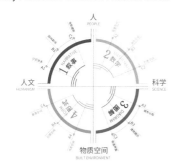

本科二年级
建筑设计基础
·刘铨　冷天
课程类型：必修
学时学分：64学时 / 4学分

Undergraduate Program 2nd Year
BASIC DESIGN OF ARCHITECTURE
· LIU Quan, LENG Tian
Type: Required Course
Study Period and Credits: 64 hours/4 credits

课题内容
认知与表达
教学目标
本课程是建筑学专业本科生的专业通识基础课程。本课程的任务主要是一方面让新生从专业的角度认知与实体建筑相关的基本知识，如主要建筑构件与材料、基本构造原理、空间尺度、建筑环境等知识；另一方面通过学习运用建筑学的专业表达方法来更好地掌握这些建筑基本知识，为今后深入的专业学习奠定基础。
教学内容
1.认知建筑
（1）立面局部测绘
（2）建筑平、剖面测绘
（3）建筑构件测绘
2.认知图示
（1）单体建筑图示认知
（2）建筑构件图示认知
3.认知环境
（1）街道空间认知
（2）建筑肌理类型认知
（3）地形与植被认知
4.专业建筑表达
（1）建筑图纸表达
（2）建筑模型表达
（3）环境分析图表达

Subject Content
Cognition and Presentation
Training Objective
The course is the basic course of general professional knowledge for undergraduates of architecture. The task of the course is, on the one hand, allow students to cognize basic knowledge about physical building from an professional perspective, such as main building components and materials, basic constructional principles, spatial dimensions, and building environment etc.; and on the other hand, to better master such basic architectural knowledge through studying application of professional presentation method of architecture, and to lay down solid foundation for future in-depth study of professional knowledge.
Teaching content
1. Cognizing building
(1) Surveying and drawing of partial elevation
(2) Surveying and drawing plans, profiles of building
(3) Surveying and drawing building components
2. Cognizing drawings
(1) Cognition to drawings of individual building
(2) Cognition to drawings of building components
3.Cognizing environment
(1) Cognition to street space
(2) Cognition to types of building texture
(3) Cognition to terrain and vegetation
4. Professional architectural presentation
(1) Presentation with architectural drawings
(2) Presentation with architectural models
(3) Presentation with environmental analysis charts

本科二年级
建筑设计（一）：独立居住空间设计
· 刘铨　冷天　黄华青
课程类型：必修
学时学分：64学时 / 4学分

Undergraduate Program 2nd Year
ARCHITECTURAL DESIGN 1: INDEPENDENT LIVING SPACE DESIGN
· LIU Quan, LENG Tian, HUANG Huaqing
Type: Required Course
Study Period and Credits: 64 hours/4 credits

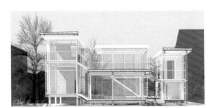

教学内容
　　本次练习的主要任务是，综合运用在建筑设计基础课程中的知识点，初步体验一个小型独立居住空间的设计过程。训练的重点在于内部空间的整合性设计，同时希望学生在设计学习开始之初，能够主动去关注场地与界面、功能与空间、流线与出入口、尺度与感知等设计要素之间的紧密关系。
教学要点
　　1. 场地与界面：场地从外部限定了建筑空间的生成条件。作为第一个设计训练，教案对场地环境条件做了简化限定，主要是要求学生从场地原有界面出发来考虑新建筑的形体、布局及其最终的空间视觉感受。
　　2. 功能与空间：使用者的不同功能需求是建筑空间生成的主要动因，也是建筑设计要解决的基本问题。本次设计的建筑功能为小型家庭独立式住宅并附设有书房功能，家庭主要成员包括一对年轻夫妇和1位未成年小孩（7岁左右），新建建筑面积160~200 m²，建筑高度≤8 m。
　　3. 流线与出入口：建筑内部各功能空间需要合理的水平、垂直交通来相互沟通与联系；建筑的内部空间需要考虑与场地周边环境条件的合理衔接。
　　4. 尺度与感知：建筑内部的空间是供人来使用的，因此建筑中的各功能空间的尺度，都必须以人体作为基本的参照和考量，并结合人体的各种行为活动方式，来确定合理的建筑空间尺寸。

Teaching content
The main task of this exercise is to comprehensively apply the knowledge points in the basic course of architectural design, and preliminarily experience the design process of a small independent living space. The training focuses on the integrated design of internal space. At the beginning of design study, students are expected to actively pay attention to the close relationship between site and interface, function and space, circulation and access, scale and perception.
Teaching essential
1. Site and interface: The site defines the generation condition of building space from the outside.
2. Function and space: Different functional requirements of users are the main motivation for the generation of building space and also the basic problem to be solved in architectural design.
3. Circulation and access: Each functional space inside the building shall be mutually communicated and connected by rational horizontal and vertical traffic; the internal space of the building shall be properly connected with the surrounding environmental conditions of the site.
4. Scale and perception: The space inside the building is for use. Therefore, the scale of each functional space in the building must take human body as the basic reference and consideration, and combine with various behavioral and activity modes of human body to confirm rational architectural space size.

本科二年级
建筑设计（二）：文怀恩旧居加建设计
· 刘铨　冷天　黄华青
课程类型：必修
学时学分：64学时 / 4学分

Undergraduate Program 2nd Year
ARCHITECTURAL DESIGN 2: DESIGN FOR EXTENSION OF WEN HUAIEN'S FORMER RESIDENCE
· LENG Tian, LIU Quan, HUANG Huaqing
Type: Required Course
Study Period and Credits: 64 hours/4 credits

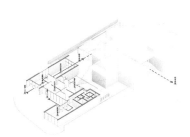

教学内容
　　1.功能要求：根据文怀恩与金陵大学的历史，设计一个展示纪念馆。老建筑由于缺少大空间，因此需要加设一个较大的灵活空间。建成后老建筑用作固定展陈和办公，新建筑则作为临时展览、研讨交流、会议茶歇等可以灵活使用的空间。同时由于其位于校园核心地带，新增建筑内拟设一个小型咖啡厅，服务学校教职工及日常参观人群。新建建筑总面积不少于150 m²，建筑高度≤8 m（檐口高度，不包括女儿墙），新建筑以一层为主，可局部夹层。在场地内还应考虑一处与展览主题相关的纪念性空间。新老建筑应作为一个整体考虑其参观流线，但新建建筑也应考虑其相对独立性，在老建筑闭馆时可独立使用。
　　2.场地环境：现状建筑和场地内各项要素既是限制，又是形成新建筑体量的基本条件。
　　3.空间限定要素与视觉关系的组织：本次训练需要通过空间限定要素与身体感知关系的组织，塑造出相应的室内外展陈与纪念性空间，连接文怀恩故居历史记忆与现实需求，创造性地再现该场所的人文内涵。

Teaching content
1.Functional requirements:Design an Exhibition Hall based on the history of Wen Huaien and Jinling University. Due to lack of large space, the old building should be attached with a larger flexible space. The completed old building will be used for exhibition and office, and the new building will be used for temporary exhibition, seminars and exchanges, conferences and tea breaks. At the same time, due to its location in the core area of the campus, it is planned to set a small coffee shop in the new building, serving the teaching and administrative staff and daily visitors. The total area of the new building should not be less than 150 m², with the building height ≤8m (the height of cornice, excluding the parapet). The new building is mainly one storey, with an interlayer in some parts. A commemorative space related to the theme of exhibition should be arranged. The visiting flow should be determined through taking the new and old buildings as a whole, but the independence of the new building should be considered, so as to ensure that it can be used when the old building is closed.
2.Site environment: The existing building and various elements of the site are limitations and basic conditions for the formation of new building volume.
3.Organization of the spatial limitation elements and visual relationships: In this training, the students should create corresponding indoor and outdoor exhibition and commemorative space based on the organization of the relationship between spatial limitation elements and body perception, so as to link the historical memory to actual needs, and creatively reproduce the humanistic connotation of the place.

本科三年级
建筑设计（三）：幼儿园设计
·童滋雨　华晓宁　窦平平
课程类型：必修
学时学分：72 学时 / 4 学分

Undergraduate Program 3rd Year
ARCHITECTURAL DESIGN 3: THE KINDERGARTEN DESIGN
• TONG Ziyu, HUA Xiaoning, DOU Pingping
Type: Required Course
Study Period and Credits: 72 hours/4 credits

教学目标
此课程训练解决建筑设计中的一类典型问题：标准空间单元的重复和组合。建筑一般都是多个空间的组合，其中一类比较特殊的建筑，其主体是通过一些相同或相似的标准空间单元重复而成，这种连续且有规律的重复，很容易表现出一种韵律和节奏感。对这类建筑的设计练习，可以帮助学生了解并熟悉空间组合中的重复、韵律、节奏、变化等操作手法。

教学内容
某幼儿园，用地面积约 7200 m²。拟设托班、小班、中班、大班各 3 个，共计 12 个班，每班人数为 25 人。使用面积约为 2600 m²。高度不超过 3 层。

Teaching objective
The course training aims to solve a typical problem in architectural design: The repetition and combination of standard space units. A building is generally a combination of multiple space units. There is a type of special buildings, whose principal parts are composed by repetition of some identical or similar standard space units. Such a continuous and regular repetition can easily reflect a sense of rhythm. The exercises in design of such buildings can help the students to understand and familiarize the manipulation techniques for spatial combination such as repetition, rhythm and variation.

Teaching content
Design a kindergarten, with the land area of about 7200 m². It is planned to set up three nursery classes, three younger classes, three middle classes, and three top classes, 12 ones in total, with 25 children each. The usable area is about 2 600 m² and its height should not exceed 3 storeys.

本科三年级
建筑设计（四）：书画家纪念馆
·华晓宁　窦平平　孟宪川
课程类型：必修
学时学分：72 学时 / 4 学分

Undergraduate Program 3rd Year
ARCHITECTURAL DESIGN 4: MUSEUM FOR ARTIST • HUA Xiaoning, DOU Pingping, Meng Xianchuan
Type: Required Course
Study Period and Credits: 72 hours/4 credits

教学目标
本课程主题是"空间"，学习建筑空间组织的技巧和方法，训练空间的操作与表达。
空间问题是建筑学的基本问题。课题基于复杂空间组织的训练和学习，从空间秩序入手，安排大空间与小空间、独立空间与重复空间，区分公共与私密空间、服务与被服务空间、开放与封闭空间。同时，充分考虑人在空间中的行为、空间感受，尝试以空间为手段表达特定的意义和氛围，最终形成一个完整的设计。

教学内容
作为十朝古都，南京历来是人文荟萃、名家辈出之地。故拟在长江路历史文化街区建造一座知名书画家的纪念馆，以促进社会文化事业的发展。
基地位于江苏美术馆老馆北侧，碑亭巷与石婆婆庵交叉口西南角，用地面积约 4100 m²。
新建筑应具备以下功能：
1）展示书画家生平、事迹、作品、影响等；
2）收藏与该书画家相关的资料、档案、文物、作品等；
3）研究与该书画家相关的历史、理论、创作等；
4）艺术普及和社会教育功能，如举办艺术讲座、研讨会、沙龙等。

Teaching objective
This course, themed in "space", intends to enable the students to learn architectural space organization skills and methods, and train them in terms of the operation and expression of space.
The problem of space is a basic problem of architecture. Based on the training and learning of complex spatial organization, this course focuses on the spatial order, and arranges the large space and small space, independent space and repetitive space, and distinguishes the public space from private space, serving space from service space, and open space from closed space. At the same time, it also fully considers people's behaviors and space perception in space, and tries to express specific meaning and atmosphere by space, thus forming a complete design.

Teaching content
Nanjing, as an ancient capital of ten dynasties, has been a place with numerous celebrities. Therefore, it is planned to build a memorial of a famous calligrapher & painter in the historical and cultural block of Changjiang Road, so as to promote the development of social and cultural undertakings.
The base is located at the north side of the old building of Jiangsu Art Museum, namely the southwest corner of Beiting Lane and Shipopo Nunnery, with the land area of about 4100 m².
The new building should have the following functions:
1) Show the life, story, works and influence of the calligrapher & painter;
2) Collect the materials, files, cultural relics, and works related to the calligrapher & painter;
3) Study the history, theory, creation etc. relevant to the calligrapher & painter;
4) Have the function of art popularization and social education, such as holding art lectures, seminars, and salons.

本科三年级

建筑设计（五）：大学生健身中心改扩建设计·傅筱 王铠 钟华颖

课程类型：必修

学时学分：64 学时 / 4 学分

Undergraduate Program 3rd Year
ARCHITECTURAL DESIGN 5: RENOVATION AND EXTENSION DESIGN OF COLLEGE STUDENT FITNESS CENTER · FU XOAO, WANG Kai, ZHONG Huaying,

Type: Required Course

Study Period and Credits: 64 hours/4 credits

教学目标

本课题以大学生健身中心为训练载体，学习并掌握中小跨建筑的基本设计原理，掌握基本的结构类型与建筑形式空间之间的逻辑关系，培养建筑结构、建筑空间与建筑设计的协调能力。

教学内容

本项目拟在南京大学鼓楼校区体育馆基地处改扩建大学生健身中心，以服务于南京大学师生，可适当考虑对周边居民的服务。根据基地条件、功能使用进行建筑和场地设计。基地用地面积 9000 m²。现状基地由一座体育馆和一座游泳馆（吕志和馆）组成，设计需保留体育馆，拆除现状游泳馆并重新设计一座大学生健身中心，其建筑总面积约 4300 m²。建筑高度控制在 24 m 以下，注意场地东西向高差，场地下挖不得超过一层，深度不超过 4.5 m。

Teaching objective

This subject takes college student fitness center as the training carrier, in which students learn and master the basic design principles of small and medium-span buildings, master the logical relationship between the basic structural type and architectural form space, and cultivate the coordination ability of architectural structure, architectural space and architectural design.

Teaching content

In this project, it is planned to renovate and expand the fitness center at the base of the gymnasium of Gulou Campus of Nanjing University, so as to serve the teachers and students of Nanjing University, and the service to surrounding residents can be properly considered. Architecture and site are designed according to the base condition and functional use. The base land area is 9000 m². The current base consists of a gymnasium and a swimming pool (Lu Zhihe Hall). The gymnasium shall be retained, the current swimming pool shall be dismantled and a new college student fitness center shall be designed, with the total building area of about 4300 m². The height of the building shall be controlled below 24 m. Students shall pay attention to the elevation difference between the east and west of the site, dig no more than one floor below the site and the depth shall not exceed 4.5 m.

本科三年级

建筑设计（六）：社区文化艺术中心设计

·张雷 钟华颖 王铠

课程类型：必修

学时学分：64 学时 / 4 学分

Undergraduate Program 3th Year
ARCHITECTURAL DESIGN 6: COMMUNITY CULTURE AND ART CENTER DESIGN
· ZHANG Lei, WANG Kai, ZHONG Huaying

Type: Required Course

Study Period and Credits: 64 hours/4 credits

教学目标

本课题以社区文化艺术中心为训练载体，学习并掌握综合功能建筑基本设计原理，理解老城区街区式建筑与城市环境的逻辑关系，培养建筑结构、建筑空间与建筑功能的综合组织能力。

教学内容

本项目拟在百子亭风貌区基地处新建社区文化中心，总建筑面积约 8000 m²，项目不仅为周边居民文化基础设施服务，同时也期望成为复兴老城的街区活力的文化地标。根据基地条件、功能使用进行建筑和场地设计。总用地详见附图，基地用地面积 4600 m²。

Teaching objective

This subject takes the community culture and art center as the training carrier, in which students learn and master the basic design principles of comprehensive functional architecture, understand the logical relationship between the block style architecture in the old urban area and the urban environment, and cultivate the comprehensive organizational ability of architectural structure, architectural space and architectural function.

Teaching content

This project plans to build a new community cultural center at the base in the Baiziting Scenic Area, with the total building area of about 8000 m². It will serve surrounding residents as the cultural infrastructure, and will also become a cultural landmark for revitalizing the neighborhood of the old town. The architectural and site design should be carried out according to base conditions and functions. The overall land use is shown in the attached drawing, and the base covers an area of 4600 m².

本科四年级
建筑设计（七）：高层办公楼设计
· 吉国华　胡友培　尹航
课程类型：必修
学时学分：64学时 / 4学分

Undergraduate Program 4th Year
ARCHITECTURAL DESIGN 7: DESIGN OF HIGH-RISE OFFICE BUILDING
· JI Guohua, HU Youpei, YIN Hang
Type: Required Course
Study Period and Credits: 64 hours/4 credits

教学内容
生态性能驱动的办公建筑设计涉及城市、空间、形体、环境、能耗、结构、设备、材料、消防等方面内容，是一项较复杂与综合的任务。有效的空间组织、适应性形体、交互性表皮以及性能化构造设计等策略，对建筑室内外环境的生态性能起着决定性的作用。

教学目标
本课题教学重点和目标是帮助学生理解、消化以上涉及各方面知识，提高综合运用并创造性解决问题的技能，学习并运用生态性能模拟分析软件，以生态性能驱动建筑设计。

Teaching content
The design of office buildings driven by eco-performance involves the aspects of city, space, form, environment, energy consumption, structure, equipment, materials, and fire protection etc.. It is a complex and comprehensive task. The strategies such as effective spatial organization, adaptive shape, interactive surface, and performance-based structural design play a decisive role in ecological performance of indoor and outdoor environment.

Teaching objective
This course intends to help the students to understand and digest the knowledge of various aspects, improve comprehensive application and creative problem solving skills, learn and use the ecological performance simulation analysis software, and drive architectural design with ecological performance.

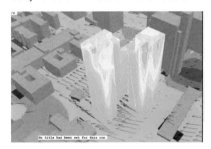

本科四年级
建筑设计（八）：城市设计
· 童滋雨　唐莲　尤伟
课程类型：必修
学时学分：64学时 / 4学分

Undergraduate Program 4th Year
ARCHITECTURAL DESIGN 8: URBAN DESIGN · TONG Ziyu, TANG Lian, YOU Wei
Type: Required Course
Study Period and Credits: 64 hours/4 credits

教学内容
经历了三十多年快速城市化，我国沿海发达城市终于不得不结束轻松的扩张，开始面对土地终会枯竭的事实。基于我们身处的城市，探讨借由改变土地使用性质、增大建设密度和改善交通组织等方式来提高土地使用效率，既是我们城市设计相关学术探索的重要方向，同时也是我们基于教研平台持续进行的系列教学实验之一。在此背景下，本课程基于真实地块的设计训练，认知高密度城市空间形态的真正含义，了解城市建筑角色和城市物质空间的本质和效能，初步掌握高质量城市空间和城市建筑组合之间的关系，同时，进一步深化空间设计的技能、方法与绘图能力。

教学目标
在设计开展之前，首先需要建立起对城市空间形态的准确理解，为此城市空间形态的认知训练与设计训练同等重要；另外，城市设计图示不同于常规的建筑学图示，也是训练的核心内容。

Teaching content
After more than 30 years of rapid urbanization, the developed coastal cities in China have finally had to end their easy expansion and begun to face the fact that land will eventually be exhausted. Based on the city we live in, discussing to improve land use efficiency by changing land use nature, increasing building density and improving transportation organization is not only an important direction of academic exploration related to urban design, but also one of the series of teaching experiments we continue on the teaching and research platform. Under this background, this course bases on real plot design training, aims to make students understand the real meaning of high-density urban space form, know the essence and effect of the role of the city building and urban physical space, preliminarily master the relationship between high quality urban space and urban architecture combination, at the same time, further deepen the skill, method and drawing ability in space design .

Teaching objective
To the senior students who contact urban design for the first time, they should first establish an accurate understanding of urban spatial form before starting design. Therefore, the cognitive training of urban spatial form is equally important as the design training. In addition, urban design diagrams are different from conventional architectural diagrams, which are also the core of the training.

本科四年级
毕业设计
· 吉国华 李清朋
课程类型：必修
学时学分：1学期 /0.75 学分

Undergraduate Program 4th Year
GRADUATION PROJECT
• JI Guohua, LI Qingpeng
Type: Required Course
Study Period and Credits: 1 term /0.75 credit

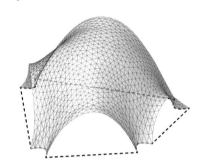

课题内容
数字化设计与建造
教学内容
本课题以"基于力学生形的数字化设计与建造"为主题，要求学生在学校自选环境中设计一处用地面积 4 m×4 m，遮盖面积为 10 m² 左右的建筑空间，以满足师生停留、休憩、交流的功能需求。课题通过实物模型制作来不断探索设计问题，用数字化的方法研究和解决问题，最终通过数控加工的方式来实现具有真实细节的构筑物。

教学目标
基于建筑数字化技术，本毕业设计涵盖案例分析、设计研究以及建造实践三个部分，建立基于力学生形的设计方法，解决数字化设计与实际建造的真实问题，完成从形态设计到数字化建造的全过程。整个课程以结构性能为形态设计的出发点，协同思考形式美学与建造逻辑的关系，培养学生在建筑设计阶段主动考虑结构逻辑的能力，在建筑形式创新和结构逻辑之间寻求统一。

Subject Content
Digital Design and Building
Teaching content
Themed in "the digital design and construction based on the formation according to mechanics", this course requires each student designing the architectural space with the site area of 4 m×4 m, and covering area of 10 m² in the school, to enable the teachers and students to stay, rest and communicate. It continues to explore design issues through developing physical models, study and solve problems by a digital method, and finally construct the structure with real details by CNC.

Teaching objective
Based on the digital technology of architecture, this graduation project covers case analysis, design research and construction practice. It intends to establish a design method based on the principle of formation according to mechanics, so as to solve the problems of digital design and actual construction, thus completing the whole process from morphological design to digital construction. The whole course takes structural performance as the starting point of form design, and coordinately considers the relationship between formal aesthetics and construction logic, to cultivate the students' ability to actively consider structural logic at the stage of architectural design, and seek unity between architectural form innovation and structural logic.

本科四年级
毕业设计
· 黄华青
课程类型：必修
学时学分：1学期 /0.75 学分

Undergraduate Program 4th Year
GRADUATION PROJECT • HUANG Huaqing
Type: Required course
Study Period and Credits: 1 term /0.75 credit

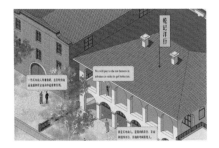

课题内容
空间作为能动者：海上茶路的技术传播与东南亚近代聚落景观的塑造
教学内容
本课题在"一带一路"、聚落文化、近代遗产等视野下，以 18 世纪中叶至 20 世纪初兴盛的海上茶叶贸易之路沿线聚落为载体，探讨茶叶技术的传播与东亚、南亚等地近现代聚落景观的形成变迁之间的激烈互动。

教学目标
本课题基于史料和田野调查，从中外聚落形态和建筑类型比较出发，为东亚与南亚近代聚落景观的空间形塑寻求以"人"为基准的线索，探索中国传统聚落和乡土聚落的海外源头，也试图为"一带一路"等多边发展倡议提供历史和理论支撑。

本课题涉及建筑学、聚落研究、建筑遗产、民族志、移民史、贸易史、技术史等多领域话题，以史料搜集、田野调查、聚落形态及建筑类型研究为手段，成果以毕业论文形式呈现。

Subject Content
Space as Agent: Transmission of Tea Technology along the Maritime Tea Road and the Shaping of Modern Settlement Landscape in East and South Asia
Teaching content
Under the background of the "Belt and Road", settlement culture, and modern heritage, the dissemination of tea technology, and the intense interaction with formation and changes of modern settlement landscapes in East Asia and South Asia would be discussed with the settlement formed along the maritime tea road prospered from the middle of the 18th century to the beginning of the 20th century as the carrier.

Teaching Objective
In this course, the settlement patterns and building types would be compared between China and foreign countries based on historical data and field research, to seek the clues based on "human" for space of modern settlements in East and South Asia, and explore the overseas sources of Chinese traditional settlements and rural settlements, thus providing the historical and theoretical support for multilateral development Initiatives of the "Belt and Road".
This topic involves architecture, settlement research, architecture heritage, ethnography, immigration history, trade history, and technology history, and it is researched by means of historical data collection, field investigation, and research of settlement pattern and architecture type. The achievements should be presented in the form of graduation thesis.

研究生一年级
建筑设计研究（一）：基本设计
·傅筱
课程类型：必修
学时学分：40 学时 / 2 学分

Graduate Program 1st Year
DESIGN STUDIO 1: DESIGN BASICS
· FU Xiao
Type: Required Course
Study Period and Credits: 40 hours/2 credits

课程内容
住宅兼工作室设计
教学目标
课程从"场地、空间、功能、经济性"等建筑的基本问题出发，通过宅基地住宅设计、训练学生对建筑逻辑性的认知，并让学生理解有品质的设计是以基本问题为基础的。
研究主题
设计的逻辑思维
教学内容
在 A、B 两块宅基地内任选一块进行住宅设计。

Course content
House and Studio
Teaching objective
The course starts from fundamental issues of architecture such as "site, space, function, and economy", aims to train students to cognize architectural logics, and allow them to understand that quality design is based on such fundamental issues.
Research subject
Logical Thinking of Design
Teaching content
Select one from two homesteads A and B, conduct housing design.

研究生一年级
建筑设计研究（二）：概念设计
·鲁安东
课程类型：必修
学时学分：40 学时 / 2 学分

Graduate Program 1st Year
DESIGN STUDIO 2: CONCEPTUAL DESIGN
· LU Andong
Type: Required Course
Study Period and Credits: 40 hours/2 credits

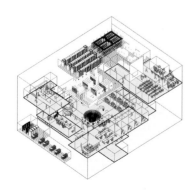

课程内容
作为空间实验室的建筑学：文创能力中心设计研究
教学内容
本课程将探索一种全新的建筑类型，即作为"空间实验室"的建筑，它植根于当代的社会经济、媒体技术和城市空间实践，兼容了物质空间与虚拟空间。本课程将针对文创这一新兴产业开展系统研究，以紫金·创联盟这一产、学、研、资、媒联动机制为切入点，分析当代文创的空间特性和形式创新，并为紫金·创联盟设计一个文创能力中心，使其成为驱动江苏省文化创新的空间实验室。
教学目标
本课程将尝试回答下列问题：
1. 什么是"空间实验室"？它对建筑有哪些新要求？建筑如何回应这些要求？
2. 文创产业如何构成？有哪些主要活动及空间需求？产、学、研、资、媒等要素在其中的各自角色与联动方式。
3. 紫金·创联盟如何作为江苏文创产业的支持平台？如何通过空间设计使其成为有效的能力中心？
4. 完成文创能力中心改造方案设计。

Course content
Architecture as Spatial Laboratory: Design Research on Competence Center for Cultural Innovatlon
Teaching content
This course will explore a new type of architecture, the architecture as "spatial laboratory", which is rooted in contemporary social economy, media technology and urban space practice, and is compatible with physical space and virtual space. It will carry out systematic research based on the emerging industry of the cultural and creative industry, and analyze the spatial characteristics and form innovation of contemporary cultural and creative industry with Purple Creative Union as the entry point; in addition, it will also design a cultural and creative competence center for Purple Creative Union, and make it a spatial laboratory driving cultural innovation in Jiangsu Province.
Teaching objective
This course will try to answer the following questions:
1. What is a "spatial laboratory"? What are the new requirements for architecture? How does the architecture respond to these requirements?
2. What is the structure of the cultural and creative industry? What are the main activities and space requirements? What are the roles and linkage methods of industry, university, institute, capital, and media?
3. How does Purple Creative Union serve as a support platform for the cultural and creative industry in Jiangsu Province? How to make it an effective competence center through space design?
4. Complete the design of transformation of competence center for cultural Innovation.

研究生二年级
建筑设计研究（三）：国际和平城市空间设计研究
· 鲁安东

课程类型：必修
学时学分：40学时 / 2学分

Graduate Program 2nd Year
DESIGN STUDIO 3: DESIGN RESEARCH ON THE INTERNATIONAL CITY OF PEACE
· LU Andong
Type: Required Course
Study Period and Credits: 40 hours/2 credits

教学内容

本次课程在"南京国际和平城市建设方案"框架下，针对"内容导向的城市设计"进行研究。在城市存量更新的背景下，内容导向的城市设计正变得日益重要，它对于城市的品牌塑造和资源整合、对于城市的文化建设和社区营造都是重要途径。本次课程将探索不同形式的记忆载体如何作为空间设计和场所设计的核心，探索物质空间设计与其他媒介、技术干预的整合可能性。在此基础上，对南京国际和平城市建设提供前瞻性的可行方案。

Teaching content

Under the framework of the "Construction Scheme of Nanjing International City of Peace", the research will be carried out on the "content-oriented urban design". The content-oriented urban design has been more and more important under the background of urban inventory renewal, and it is also an important method for brand building, resource integration, urban cultural construction and community building. This course will explore how different forms of memory carriers act as the core of space design and place design, and also explore the possibility of integration of physical space design with other media and technological interventions. On such a basis, it will provide a forward-looking and feasible scheme for the construction of Nanjing international city of peace.

研究生二年级
建筑设计研究（三）：策划—规划—建筑—景观整合设计训练
· 周凌

课程类型：必修
学时学分：40学时 / 2学分

Graduate Program 2nd Year
DESIGN STUDIO 3: PLOTTING, PLANNING, ARCHITECTURE AND LANDSCAPE INTEGRATED DESIGN TRAINING · ZHOU Ling
Type: Required Course
Study Period and Credits: 40 hours/2 credits

教学内容

通过一个具体实际项目的设计训练，掌握策划、规划、建筑、景观等多专业的设计技能与相关知识。具体涉及产业策划、旅游策划、产业规划、环境规划、交通规划等策划与规划方面的知识，涉及建筑设计的基本的结构构造知识、功能组织、形体设计技能。通过课程设计，掌握规划建筑专业各种相关的知识和技能，同时训练各种形式的表达。

教学进度

第1周：现场调研，现状分析。
第2周：策划方案与概念规划。
第3周：完成规划。
第4周：建筑概念设计。
第5~6周：建筑设计。
第7~8周：成果表达制作。

Teaching content

Help the students to master design skills and knowledge of plotting, planning, architecture and landscape through design training of a specific project, which specifically involves the knowledge of industrial plotting, tourism plotting, industrial planning, environmental planning, and transportation planning, as well as the basic structural knowledge, functional organization, and shape design skills in architectural design. This course will train various forms of expression in addition to the help for students to master various knowledge and skills of planning and construction through course design.

Teaching process

Week 1: Site investigation, and current situation analysis.
Week 2: Plotting scheme and conceptual planning.
Week 3: Completion of planning.
Week 4: Architecture concept design.
Week 5~6: Architectural design.
Week 7~8: Preparation for achievement expression.

研究生二年级
建筑设计研究（三）：内边缘 2.0：城市交通廊道综合再利用
· 胡友培
课程类型：必修
学时学分：40 学时 / 2 学分

Graduate Program 2nd Year
DESIGN STUDIO 3: INNER EDGE 2.0: COMPREHENSIVE REUSE OF URBAN TRAFFIC CORRIDORS · HU Youpei
Type: Required Course
Study Period and Credits: 40 hours/2 credits

任务与要求

场地研究：课程在南京都市区内选择四处典型的交通廊道地带，作为设计研究的场地。研究场地在城市化进程中的变迁，阐明基础设施在其中的主导地位，并分析呈现各种伴生性的城市问题。

场地愿景：从城市发展的全局视角，研判场地的价值与潜力，为场地建构新的身份，提出新的愿景。制定场地的功能计划，与策略性的项目介入，探索场地再利用的可能与路径。

原形设计：课程将以原形设计为主要的设计工具。通过对建筑、基础设施、景观以及它们的混杂，开展原形设计，赋予场地某种适宜的、具有想象力的空间形态。

Tasks and requirements

Site study: Select four typical traffic corridors in Nanjing metropolitan area as the site for design and research. study the changes of the site in the process of urbanization, clarify the dominant position of infrastructure in the process of urbanization, and analyze various accompanying urban problems.

Site vision: Study and judge the value and potential of the site from the overall perspective of urban development, construct a new identity for the site, put forward a new vision, formulate the functional plan of the site and strategic project intervention, and explore the possibility and path of site reuse.

Prototype design: Take prototype design as the main design tool, and give the site a suitable and imaginative space form through prototype design of architecture, infrastructure, landscape and their mixture.

研究生国际教学交流计划
创新项目：南京周边空间环境再生
· 凯瑞·希瑞斯
课程类型：必修
学时学分：40 学时 / 2 学分

The International Postgraduate Teaching Programme
INNOVATION PROJECTS: REGENERATIVE ENVIRONMENTS FOR PERIPHERAL NANJING · Cary Siress
Type: Required Course
Study Period and Credits: 40 hours/2 credits

教学内容

1. 介绍"创新项目"和"跨行业订单"的概念，并结合研讨会课程中介绍的理论材料，提供这些概念的论述背景；

2. 介绍各种城市干预尺度、场地分析模式、场景建模和时间分段的技术，以及社会和空间领域的概念；

3. 介绍绘图、图表、场景建模和模型制作的多种技术，这些技术均为空间表达和设计思想交流的创新手段；

4. 介绍与其他个人项目提案相关的建筑、基础设施和景观设计方面的参考资料，并向学生提供选定参考项目的概念和实用框架；

5. 介绍一系列在专业公共论坛上展示城市研究成果的方法，学生将熟练掌握批判性设计思维，以专业的形式管理其工作，并以英语进行公开展示。

Teaching content

1. Students will be introduced to the notion of "heterotypes" as well as "conglomerate orders" and will be given a discursive background of these concepts in conjunction with the theoretical material introduced in the seminar course;
2. Students will be introduced to various urban scales of intervention, modes of site analysis, techniques for scenario modeling and time phasing, as well as the notion of social and spatial territoriality;
3. Students will be introduced to multiple techniques of drawing, diagramming, scenario modeling, and model making as innovative means of spatial expression and communication of design ideas;
4. Students will be introduced to references from architecture, infrastructure, and landscape design relevant to their individual project propositions and will be provided with the conceptual and pragmatic framework for the selected reference projects;
5. Students will be introduced to a range of techniques for professionally presenting urban research findings in a public forum and will be skilled in critical design thinking, curating their work in a professional format, and making public presentations in English.

研究生国际教学交流计划
让自然融入我们的建筑之中
· 伊斯梅尔·多明格兹
课程类型：必修
学时学分：40 学时 / 2 学分

The International Postgraduate Teaching Programme
ALLOWING NATURE TO BE A PART OF OUR ARCHITECTURE · Ismael Domínguez
Type: Required Course
Study Period and Credits: 40 hours/2 credits

教学内容

本课程目的是设计出顾及环境条件以及城市、社会和景观方面的当代建筑。为此，本课程通过四个基础来开发，即案例研究、环境工具、设计方法和策略以及规划行动。

提供"将环境因素融入建筑中的不同方式"的列表，每个学生团队从中选择一个案例研究。将通过双重过滤、符合建筑和可持续的方法对案例进行整体分析。

从该分析中，学生将通过归纳学习不同的策略和设计方法来面对他们的项目，一方面考虑当地和全球的文化价值，另一方面考虑环境质量和高效率表现。

讲座课程将采用每周讲座的方式，主题范围的重点是：为开发高质量和高性能的建筑和城市设计，项目考虑自然流动和环境的重要性和影响。

Teaching content

This course aims to design contemporary architecture taking into account the environmental conditions as well as urban, social and landscape aspects. For that purpose, the course is developed through four bases: case studies, environmental tools, designing methods and strategies, and projecting action.

Each student team will choose a case study among a list representative of "different ways to integrate environmental aspects into the architecture". They will be analyzed integrally with a double filter, architectural and sustainable approach.

From this analysis, students will learn by induction different strategies and design methods for facing their projects considering in one hand local and global cultural values, and in the other environmental quality and high efficiency performance.

The lecture course will introduce along weekly lectures, a scope of themes pointing the importance and consequences of permitting natural fluxes and environment to be an integral part of the project, for making high quality and performance architecture and urban design.

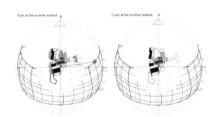

研究生设计工作坊
古卫城墙的未来可能
· 刘宇扬
课程类型：选修
学时学分：18 学时 / 1 学分

Graduate Design Workshop
THE ANCIENT WALLED CITY AND ITS FUTURE POSSIBILITIES · LIU Yuyang
Type: Elective Course
Study Period and Credits: 18 hours/1 credit

教学内容

意大利建筑师奥尔多·罗西关于城市历史、碎片和记忆的出版论述和实践作品，影响了1970—1980年代开始的一整代现当代欧洲建筑师群体。参与的研究生同学们可借此工作坊契机，通过课前阅读和课间讨论，结合永宁的城市历史及设计课题，对罗西的理论和实践形成初步认识并获得新的启发。

在为期两周的工作坊中，同学们将通过理论研讨的过程取得一系列的概念关键词，并将概念关键词转化为图像与空间操作。我将根据工作坊惯例和同学需求，合理安排线上讨论及评图。同学们可结合文字、工作草图、模型、各类表现图纸以及其他辅助视觉表现手段如虚拟现实、电影等，形成一个最终的概念设计干预提案。终期汇报提交一篇大约1000字的文字论述，一份不少于20页的PPT文件（中英文）和一个不超过3分钟的短视频。工作坊旨在从理论/方法论作为工作切入点，通过研究到设计建立一套合理的工作方法，重点在于过程中的设计实验和方法探讨，为同学们在日后的学习与实践提供一点新的思考方向。

Teaching content

The Italian architect Aldo Rossi published several works about urban history, fragments, and memory, which affected a whole generation of modern-contemporary European architects after 1970s and 1980s. All participating graduate students can take the opportunity to form a preliminary understanding of Rossi's theory and practice and obtain new inspiration through pre-class reading and inter-class discussion in combination with the history of Yongning and the design topic.

In the two-week workshop, students will obtain a series of concept keywords through the process of theoretical discussion, and transform the concept keywords into image and spatial operation. According to the working method and students' requirements, online discussion and drawing evaluation will be arranged reasonably. Students can form a final conceptual design intervention pattern by combining words, working sketches, models, various performance drawings and other auxiliary visual expression means, such as virtual reality and film. In the final report, a text discussion of about 1000 words, a PPT document of no less than 20 pages (in Chinese and English) and a short video of no more than three minutes shall be submitted in the final report. The workshop aims to establish a set of reasonable working methods from theory / methodology through research to design, focusing on the design experiment and method discussion in the process, so as to provide new thinking direction for students' study and practice in the future.

研究生设计工作坊

力透纸"贝":日常材料的催化利用
· 朱竞翔
课程类型:选修
学时学分: 18 学时 / 1 学分

Graduate Design Workshop
FORCE PENETRATING PAPER: CATALYTIC UTILIZATION OF DAILY MATERIALS
• ZHU Jingxiang
Type: Elective Course
Study Period and Credits: 18 hours/1 credit

教学内容

课程第一阶段,探究纸张和书本的性能特征,选择合理的连接方式与几何空间分布形式,进而改善材料的多重性能,实现材料的催化利用。第二阶段,学生沿着教学讨论中阐明的方向继续深入设计,并延伸到相关产品设计,结合特定的人体部位进行可行性的操作。第三阶段,针对特定的产品设计,从材料选择、连接方式和性能提升等方面进行深化、完善设计。第四阶段,通过实验数据、照片、图纸和书面描述,记录工作过程与实践结果。

在理论讲座中,朱竞翔教授首先介绍了罗伯特·马拉尔典型的桥梁工程案例,及其对于拱形管状结构的探索与实践,以此帮助同学们加深对结构性能的理解。其次介绍了其向建筑领域进行的扩展,在薄壳形态的揭示、无梁楼盖的发明、利用平面静态去理解桁架屋顶结构三个方向进行了探索。香港中文大学的博士生翟玉琨通过建筑师的作品介绍,纸筒、纸板等材料以及插接、折叠等连接方式进行解读,并对比"1958年世博会芬兰馆"与"青海拉吾尕小学"两个项目,体现结构、建造、表皮和空间等多方面的统一的设计思想。

Teaching content

At the first stage of the course, the performance characteristics of paper and books are explored, and reasonable connection method and geometric spatial distribution forms are selected, so as to improve multiple properties of materials and realize their catalytic utilization. At the second stage, all students continue to design in depth along the direction clarified in the teaching discussion, and extend to relevant product design, and carry out feasible operation in combination with specific body parts.At the third stage, all students deepen and improved the design of specific products in terms of material selection, connection method and performance improvement. At the fourth stage, all students record the work process and practical results based on experimental data, photos, drawings and written descriptions.

In the theoretical lecture, Professor Zhu Jingxiang firstly introduced the typical bridge designed by Robert Maillart in terms of the exploration and practice of the arched tubular structure, aiming to help them to deepen the understanding of structural performance. He secondly introduced the expansion to the architectural field, and made an exploration in the reveal of thin shell shape, the invention of flat slab, and the understanding of roof truss based on plane static. Zhai Yukun, doctoral student of Chinese University of Hong Kong, interpreted the materials such as paper tubes and cardboard and connection methods such as insertion and folding based on introduction of the architect's works. In addition, he compared "Finland Pavilion at World Expo 1958" and "Qinghai Lawuga Primary School", and reflected the unified design integrating structure, construction, surface and space.

研究生设计工作坊

废墟的可能性 · 张宇星
课程类型:选修
学时学分: 18 学时 / 1 学分

Graduate Design Workshop
THE POSSIBILITY OF RUINS
• ZHANG Yuxing
Type: Elective Course
Study Period and Credits: 18 hours/1 credit

教学内容

在今天,废墟越来越成为被热烈讨论的话题。一方面是由于互联网虚拟空间正在成为真实物理空间的全面替代者。互联网空间的内在废墟性,使得我们的当代城市日趋废墟化。从商业街到购物中心、从中央商务区到游乐园、从零售店到街头摊贩,无不正在抵抗着虚拟世界的强烈冲击。另一方面,废墟美学也成为未来城市的一种可能性场景,因为废墟包含了时间性和自然性,也包含了去结构化与去中心化的内在力量,它们都与未来数字世界的建构逻辑直接关联。所有这些都注定了,与废墟相关的理论和实践将成为下一代建筑学需要重点关注的学术命题。

本次工作坊分为三个部分:理论教学、设计研讨、答辩评图。

课程以深圳宝安区蚝乡湖旧电厂为案例,要求学生初步接触废墟的基本概念,梳理相关案例,学会把废墟的思想理念纳入旧建筑改造的设计过程中。通过废墟意向设计拓展设计思维,为今后的旧建筑改造设计实践打下基础。更进一步,延展思考有关"废墟建筑学"的理论,特别是对废墟所包含的建筑学本体属性。这样的深度思考将有助于学生拓宽视野,未来可以进入更加广阔的学术领域。

Teaching content

The course takes the old Haoxianghu Power Plant of Bao'an District, Shenzhen as the research case; the students are asked to preliminarily understand the basic concept of ruins, sort out the related cases, and learn to incorporate the thinking of ruins into design of renovation of old buildings. They are expected to expand the design thinking based on intended design of ruins, so as to lay a foundation for renovation design of old buildings. Furthermore, the students are expected to extend the theory of "Ruins Architecture", especially the properties of buildings contained in ruins. The in-depth thinking will help the students to broaden the vision, and enable them to step into a broader academic field in the future.

建筑理论课程
ARCHITECTURAL THEORY COURSES

Undergraduate Program 2nd Year
INTRODUCTION TO ARCHITECTURE • ZHAO Chen, etc.
Type: Required Course
Study Period and Credits: 36 hours / 2 credits

Course content
1. Preliminary of architecture
Lect. 1: Architecture and design / ZHAO Chen
Lect. 2: Architecture and urbanization / DING Wowo
Lect. 3: Architecture and life / ZHANG Lei
2. Basic attribute of architecture
1) Physical attribute
Lect. 4: Physical environment of architecture / ZHAO Chen
Lect. 5: Architecture and ecological environment / WU Wei
Lect. 6: Architecture with Materialization / Wang Dandan
2) Cultural attribute
Lect. 7: Architecture and civilization, arts, aesthetic / ZHAO Chen
Lect. 8: Architecture and landscaping environment / HUA Xiaoning
Lect. 9: Environmental Intelligence in Architecture / Dou Pingping
Lect. 10: Architecture and body / LU Andong
3) Architect: profession and academy
Lect. 11: Architecture and presentation / ZHAO Chen
Lect. 12: Architecture and geometrical form / ZHOU Ling
Lect. 13: Architecture and digital technology / ZHONG Huaying
Lect. 14: Architect's professional technique and responsibility / FU Xiao

---

Undergraduate Program 3rd Year
BASIC THEORY OF ARCHITECTURAL DESIGN • ZHOU Ling
Type: Required Course
Study Period and Credits: 36 hours / 2 credits

Teaching objective
This course is a basic theory course for the undergraduate students of architecture. The main purpose of this course is to introduce the basic principles of the form and type in architectural design. Form theory contains design principles in various periods of history. Type theory discusses the design principles of different types of buildings.
Course requirement
1. Teach the key elements of the outline;
2. Enlighten students' thinking and enhance students' understanding of the theories, also the applications and project examples through analyzing examples;
3. Assist students to use the professional knowledge to analyse and solve practical problems through the discussion of examples.
Course content
1. Overview of forms and types
2. Classical architecture form language
3. Modern architecture form language
4. Contemporary architecture form language
5. Type design
6. Materials and construction
7. Technology and specifications
8. Course summary

Undergraduate Program 3rd Year
THEORY OF HOUSING DESIGN AND RESIDENTIAL PLANNING • LENG Tian, LIU Quan
Type: Required Course
Study Period and Credits: 36 hours / 2 credits

Course content
Lect. 1: Introduction of the course
Lect. 2: Development of residential buildings
Lect. 3: Design of dwelling space
Lect. 4: Dwelling space arrangements and residential building design (1)
Lect. 5: Dwelling space arrangements and residential building design (2)
Lect. 6: Structure, detail, facility and construction of residential buildings
Lect. 7: Adaptive ability of residential buildings, supporting houses
Lect. 8: Introduction of the theories of urban planning
Lect. 9: History of modern residential planning
Lect. 10: Organization of residential space
Lect. 11: Traffic system planning and design of residential area
Lect. 12: Landscape planning and design of residential area
Lect. 13: Public facilities and infrastructure system
Lect. 14: Real estate development
Lect. 15: The practice of residential planning and housing design
Lect. 16: Summary, answer questions on the test

研究生一年级
现代建筑设计基础理论
· 张雷
课程类型：必修
学时/学分：18学时/1学分

Graduate Program 1st Year
PRELIMINARIES IN MODERN ARCHITECTURAL DESIGN · ZHANG Lei
Type: Required Course
Study Period and Credits:18 hours/1 credit

教学目标
建筑可以被抽象到最基本的空间围合状态来面对它所必须解决的基本的适用问题，用最合理、最直接的空间组织和建造方式去解决问题，以普通材料和通用方法去回应复杂的使用要求，是建筑设计所应该关注的基本原则。
课程要求
1. 讲授大纲的重点内容；
2. 通过分析实例启迪学生的思维，加深学生对有关理论及其应用、工程实例等内容的理解；
3. 通过对实例的讨论，引导学生运用所学的专业理论知识，分析、解决实际问题。
课程内容
1. 基本建筑的思想
2. 基本空间的组织
3. 建筑类型的抽象与还原
4. 材料的运用与建造问题
5. 场所的形成及其意义
6. 建筑构思与设计概念

Teaching objective
The architecture can be abstracted into spatial enclosure state to encounter basic application problems which must be settled. Solving problems with most reasonable and direct spatial organization and construction mode, and responding to operating requirements with common materials and general methods are basic principles concerned by building design.
Course requirement
1. To teach key contents of syllabus;
2. To inspire students' thinking, deepen students' understanding on such contents as relevant theories and their application and engineering example through case analysis;
3. To assist students to use professional theories to analyze and solve practical problems through discussion of instances.
Course content
1. Basic architectural thoughts
2. Basic spacial organization
3. Abstraction and restoration of architectural types
4. Utilization and construction of materials
5. Formation of site and its meaning
6. Architectural conception and design concept

研究生一年级
研究方法与写作规范
· 鲁安东 胡恒 郜志
课程类型：必修
学时/学分：18学时/1学分

Graduate Program 1st Year
RESEARCH METHOD AND THESIS WRITING
· LU Andong HU Heng GAO Zhi
Type: Required Course
Study Period and Credits:18 hours/1 credit

教学目标
面向学术型硕士研究生的必修课程。它将向学生全面地介绍学术研究的特性、思维方式、常见方法以及开展学术研究必要的工作方式和写作规范。考虑到不同领域研究方法的差异，本课程的授课和作业将以专题的形式进行组织，包括建筑研究概论、设计研究、科学研究、历史理论研究4个模块。学生通过各模块的学习可以较为全面地了解建筑学科内主要的研究领域及相应的思维方式和研究方法。
课程要求
将介绍建筑学科的主要研究领域和当代研究前沿，介绍"研究"的特性、思维方式、主要任务、研究的工作架构以及什么是好的研究，帮助学生建立对"研究"的基本认识；介绍文献检索和文献综述的规范和方法；介绍常见的定量研究、定性研究和设计研究的工作方法以及相应的写作规范。
课程内容
1. 综述
2. 文献
3 科学研究及其方法
4. 科学研究及其写作规范
5. 历史理论研究及其方法
6. 历史理论研究及其写作规范
7. 设计研究及其方法
8. 城市规划理论概述

Teaching objective
It is a compulsory course to MA. It comprehensively introduces features, ways of thinking and common methods of academic research, and necessary manners of working and writing standards for launching academic research to students. Considering differences of research methods among different fields, teaching and assignment of the course will be organized in the form of special topic, including four parts: introduction to architectural study, design study, scientific study and historical theory study. Through the study of all parts, students can comprehensively understand main research fields and corresponding ways of thinking and research methods of architecture.
Course requirement
The course introduces main research fields and contemporary research frontier of architecture, features, ways of thinking and main tasks of "research", working structure of research, and definition of good research to help students form basic understanding of "research". The course also introduces standards and methods of literature retrieval and review, and working methods of common quantitative research, qualitative research and design research, and their corresponding writing standards.
Course content
1. Review
2. Literature
3. Scientific research and methods
4. Scientific research and writing standards
5. Historical theory study and methods
6. Historical theory study and writing standards
7. Design research and methods
8. Overview of urban planning theory

城市理论课程
URBAN THEORY COURSES

本科四年级
城市设计理论
· 胡友培
课程类型：必修
学时/学分：36学时/2学分

Undergraduate Program 4th Year
THEORY OF URBAN DESIGN • HU Youpei
Type: Required Course
Study Period and Credits: 36 hours / 2 credits

课程内容
第一讲：课程概述
第二讲：城市设计技术术语：城市规划相关术语，城市形态相关术语，城市交通相关术语，消防相关术语
第三讲：城市设计方法 —— 文本分析：城市设计上位规划，城市设计相关文献，文献分析方法
第四讲：城市设计方法 —— 数据分析：人口数据分析与配置，交通流量数据分析，功能分配数据分析，视线与高度数据分析，城市空间数据模型的建构
第五讲：城市设计方法 —— 城市肌理分类：城市肌理分类概述，肌理形态与建筑容量，肌理形态与开放空间，肌理形态与交通流量，城市绿地指标体系
第六讲：城市设计方法 —— 城市路网组织：城市道路结构与交通结构概述，城市路网与城市功能，城市路网与城市空间，城市路网与市政设施，城市道路断面设计
第七讲：城市设计方法 —— 城市设计表现：城市设计分析图，城市设计概念表达，城市设计成果解析图，城市设计地块深化设计表达，城市设计空间表达
第八讲：城市设计的历史与理论：城市设计的历史意义，城市设计理论的内涵
第九讲：城市路网形态：路网形态的类型和结构，路网形态与肌理，路网形态的变迁
第十讲：城市空间：城市空间的类型，城市空间结构，城市空间形态，城市空间形态的变迁
第十一讲：城市形态学：英国学派，意大利学派，法国学派，空间句法
第十二讲：城市形态的物理环境：城市形态与物理环境，城市形态与环境研究，城市形态与环境测评，城市形态与环境操作
第十三讲：景观都市主义：景观都市主义的理论、操作和范例
第十四讲：城市自组织现象及其研究：城市自组织现象的魅力与问题，城市自组织现象研究方法，典型自组织现象案例研究
第十五讲：建筑学图式理论与方法：图式理论的研究，建筑学图式的概念，图式理论的应用，作为设计工具的图式，当代城市语境中的建筑学图式理论探索
第十六讲：课程总结

Course content
Lect. 1: Introduction
Lect. 2: Technical terms: terms of urban planning, urban morphology, urban traffic and fire protection
Lect. 3: Urban design methods — documents analysis: urban planning and policies, relative documents, document analysis techniques and skills
Lect. 4: Urban design methods — data analysis: data analysis of demography, traffic flow, function distribution, visual and building height, modelling urban spatial data
Lect. 5: Urban design methods — classification of urban fabrics: introduction of urban fabrics, urban fabrics and floor area ratio, urban fabrics and open space, urban fabrics and traffic flow, criteria system of urban green space
Lect. 6: Urban design methods — organization of urban road network: introduction, urban road network and urban function, urban road network and urban space, urban road network and civic facilities, design of urban road section
Lect. 7: Urban design methods — representation skills of urban Design: mapping and analysis, conceptual diagram, analytical representation of urban design, representation of detail design, spatial representation of urban design
Lect. 8: Brief history and theories of urban design: historical meaning of urban design, connotation of urban design theories
Lect. 9: Form of urban road network: typology, structure and evolution of road network, road network and urban fabrics
Lect. 10: Urban space: typology, structure, morphology and evolution of urban space
Lect. 11: Urban morphology: Cozen School, Italian School, French School, Space Syntax Theory
Lect. 12: Physical environment of urban forms: urban forms and physical environment, environmental study, environmental evaluation and environmental operations
Lect. 13: Landscape urbanism: ideas, theories, operations and examples of landscape urbanism
Lect. 14: Researches on the phenomena of the urban self-organization: charms and problems of urban self-organization phenomena, research methodology on urban self-organization phenomena, case studies of urban self-organization phenomena
Lect. 15: Theories and methods of architectural diagram: theoretical study on diagrams, concepts of architectural diagrams, application of diagram theory, diagrams as design tools, theoretical research of architectural diagrams in contemporary urban context
Lect. 16: Summary

本科四年级
景观规划设计及其理论
· 尹航
课程类型：选修
学时/学分：36学时/2学分

Undergraduate Program 4th Year
LANDSCAPE PALNNING DESIGN AND THEORY
• YIN Hang
Type: Elective Course
Study Period and Credits: 36 hours / 2 credits

课程介绍
景观规划设计的对象包括所有的室外环境，景观与建筑的关系往往是紧密而互相影响的，这种关系在城市中表现得尤为明显。景观规划设计及理论课程希望从景观设计理念、场地设计技术和建筑周边环境塑造等方面开展课程的教学，为建筑学本科生建立更加全面的景观知识体系，并且完善建筑学本科生在建筑场地设计、总平面规划与城市设计等方面的设计能力。
本课程主要从三个方面展开：一是理念与历史：以历史的视角介绍景观学科的发展过程，让学生对景观学科有一个宏观的了解，初步理解景观设计理念的发展；二是场地与文脉：通过阐述景观规划设计与周边自然环境、地理位置、历史文脉和方案可持续性的关系，建立场地与文脉的设计思维；三是景观与建筑：通过设计方法授课、先例分析作业等方式让学生增强建筑的环境意识，了解建筑的场地设计的影响因素、一般步骤与设计方法，并通过与"建筑设计六"和"建筑设计七"的设计任务书相配合的同步课程设计训练来加强学生景观规划设计的能力。

Course description
The object of landscape planning design includes all outdoor environments; the relationship between the landscape and building is often close and interactive, which is especially obvious in a city. This course expects to carry out teaching from perspective of landscape design concept, site design technology, building's peripheral environment creation, etc., to establish a more comprehensive landscape knowledge system for the undergraduate students of architecture, and perfect their design abilities in building site design, master plane planning and urban design and so on.
This course includes three aspects:
1. Concept and history
2. Site and context
3. Landscape and building

研究生一年级
景观都市主义理论与方法
· 华晓宁
课程类型：选修
学时/学分：18学时/1学分

Graduate Program 1st Year
THEORY AND METHOD OF LANDSCAPE URBANISM ·
HUA Xiaoning
Type: Elective Course
Study Period and Credits: 18 hours / 1 credit

课程介绍
本课程作为国内首次以景观都市主义相关理论与策略为教学内容的尝试，介绍了景观都市主义思想产生的背景、缘起及其主要理论观点，并结合实例，重点分析了其在不同的场址和任务导向下发展起来的多样化的实践策略和操作性工具。

课程要求
1. 要求学生了解景观都市主义思想产生的背景、缘起和主要理念。
2. 要求学生能够初步运用景观都市主义的理念和方法分析和解决城市设计问题，从而在未来的城市设计实践中强化景观整合意识。

课程内容
第一讲：从图像到效能：景观都市实践的历史演进与当代视野
第二讲：生态效能导向的景观都市实践（一）
第三讲：生态效能导向的景观都市实践（二）
第四讲：社会效能导向的景观都市实践
第五讲：基础设施景观都市实践
第六讲：当代高密度城市中的地形学
第七讲：城市图绘与图解
第八讲：从原形到系统——AA景观都市主义

Course description
Combining relevant theories and strategies of landscape urbanism firstly in China, the course introduces the background, origin and main theoretical viewpoint of landscape urbanism, and focuses on diversified practical strategies and operational tools developed under different orientations of site and task with examples.

Course requirement
1. Students are required to understand the background, origin and main concept of landscape urbanism.
2. Students are required to preliminarily utilize the concept and method of landscape urbanism to analyze and solve the problem of urban design, so as to strengthen landscape integration consciousness in the future.

Course content
Lect. 1: From pattern to efficacy: historical revolution and contemporary view of practice of landscape urbanism
Lect. 2: Eco-efficiency-oriented practice of landscape urbanism (1)
Lect. 3: Eco-efficiency-oriented practice of landscape urbanism (2)
Lect. 4: Social efficiency-oriented practice of landscape urbanism
Lect. 5: Infrastructure practice in landscape urbanism
Lect. 6: Geomorphology in contemporary high-density cities
Lect. 7: Urban painting and diagrammatizing
Lect. 8: From prototype to system: AA landscape urbanism

研究生一年级
城市形态与设计方法论 · 丁沃沃
课程类型：必修
学时/学分：36学时/2学分

Graduate Program 1st Year
URBANISM FORM AND DESIGN METHODOLOGY · DING Wowo
Type: Required Course
Study Period and Credits: 36 hours / 2 credits

课程介绍
建筑学核心理论包括建筑学的认识论和设计方法论两大部分。建筑设计方法论主要探讨设计的认知规律、形式的逻辑、形式语言类型，以及人的行为、环境特征和建筑材料等客观规律对形式语言的选择及形式逻辑的构成策略。为此，设立了提升建筑设计方法的关于设计方法论的理论课程，作为建筑设计及其理论硕士学位的核心课程。

课程要求
1. 理解随着社会转型，城市建筑的基本概念在建筑学核心理论中的地位以及认知的视角。
2. 通过理论的研读和案例分析理解建筑形式语言的成因和逻辑，并厘清中、西不同的发展脉络。
3. 通过研究案例的解析理解建筑形式语言的操作并掌握设计研究的方法。

课程内容
第一讲：序言
第二讲：西方建筑学的基础
第三讲：中国：建筑的意义
第四讲：背景与文献研讨
第五讲：历史观与现代性
第六讲：现代城市形态演变与解析
第七讲：现代城市的"乌托邦"
第八讲：现代建筑的意义
第九讲：建筑形式的反思与探索
第十讲：建筑的量产与城市问题
第十一讲："乌托邦"的实践与反思
第十二讲：都市实践探索的理论价值
第十三讲：城市形态的研究
第十四讲：城市空间形态研究的方法
第十五讲：回归理性：建筑学方法论的新进展
第十六讲：建筑学与设计研究的意义
第十七讲：结语与研讨（一）
第十八讲：结语与研讨（二）

Course description
The core theory of architecture includes epistemology and design methodology of architecture. Architectural design methodology mainly discusses cognitive laws of design the, logic of form and types of formal language, and the choice of the formal language from objective laws such as human behaviors, environmental features and building materials, and composition strategies of formal logic. Thus, the theory course about design methodology to promote architectural design methods is established as the core course of architectural design and theory master degree.

Course requirement
1. To understand the status and cognitive perspective of basic concept of urban buildings in the core theory of architecture with the social transformation.
2. To understand the reason and logic of the architectural formal language and different development process in China and West through reading theories and case analysis.
3. To understand the operation of the architectural formal language and grasp methods of design study by analyzing study cases.

Course content
Lect. 1: Introduction
Lect. 2: Foundation of western architecture
Lect. 3: China: meaning of architecture
Lect. 4: Background and literature discussion
Lect. 5: Historicism and modernity
Lect. 6: Analysis and morphological evolution of modern city
Lect. 7: "Utopia" of modern city
Lect. 8: Meaning of modern Architecture
Lect. 9: Reflection and exploration of architectural form
Lect. 10: Mass production of buildings and urban problems
Lect. 11: Practice and reflection of "Utopia"
Lect. 12: Theoretical value of exploration on urban practice
Lect. 13: Study on urban morphology
Lect. 14: Method of urban spatial morphology study
Lect. 15: Return to rationality: new developments of methodology on architecture
Lect. 16: Meaning of architecture and design study
Lect. 17: Conclusion and discussion (1)
Lect. 18: Conclusion and discussion (2)

历史理论课程
HISTORY THEORY COURSES

本科二年级
中国建筑史（古代）
·赵辰
课程类型：必修
学时/学分：36学时/2学分

Undergraduate Program 2nd Year
HISTORY OF CHINESE ARCHITECTURE (ANCIENT)
• ZHAO Chen
Type: Required Course
Study Period and Credits: 36 hours / 2 credits

教学目标
本课程作为本科建筑学专业的历史与理论课程，目标在于培养学生的史学研究素养与对中国建筑及其历史的认识两个层面。在史学理论上，引导学生理解建筑史学这一交叉学科的多种棱面与视角，并从多种相关学科层面对学生进行基本史学研究方法的训练与指导。中国建筑史层面，培养学生对中国传统建筑的营造特征与文化背景建立构架性的认识体系。
课程内容
中国建筑史学七讲与方法论专题。七讲总体走向从微观到宏观，整体以建筑单体—建筑群体—聚落与城市—历史地理为序；从物质性到文化，建造技术—建造制度—建筑的日常性—纪念性—政治与宗教背景—美学追求。方法论专题包括建筑考古学、建筑技术史、人类学、美术史等层面。

Teaching objective
As a historical & theoretical course for undergraduate students, this course aims at two aspects of training: the basic academic capability of historical research and the understanding of Chinses architectural history. It will help students to establish a knowledge frame, that the discipline of History of Architecture as a cross-discipline, is supported and enriched by multiple neighboring disciplines and that the features and the development of Chinese Architecture roots deeply in the natural and cultural background.
Course Content
The course composes seven 4-hour lectures on Chinese Architecture and a series of lectures on methodology. The seven courses follow a route from individual to complex, from physical building to the intangible technique and to the cultural background, from technology to institution, to political and religious background, and finally to aesthetic pursuit. The special topics on methodology include building archaeology, building science and technology, anthropology, art history and so on.

本科二年级
外国建筑史（古代）·王丹丹
课程类型：必修
学时/学分：36学时/2学分

Undergraduate Program 2nd Year
HISTORY OF WORLD ARCHITECTURE (ANCIENT)
• WANG Dandan
Type: Required Course
Study Period and Credits: 36 hours / 2 credits

教学目标
本课程力图对建筑史的脉络做一个整体勾勒，使学生在掌握重要的建筑史知识点的同时，对西方建筑史在2000多年里的变迁的结构转折（不同风格的演变）有深入的理解。本课程希望学生对建筑史的发展与人类文明发展之间的密切关联有所认识。
课程内容
1. 概论 2. 希腊建筑 3. 罗马建筑 4. 中世纪建筑
5. 意大利的中世纪建筑 6. 文艺复兴 7. 巴洛克
8. 美国城市 9. 北欧浪漫主义 10. 加泰罗尼亚建筑
11. 先锋派 12. 德意志制造联盟与包豪斯
13. 苏维埃的建筑与城市 14. 1960 年代的建筑
15. 1970 年代的建筑 16. 答疑

Teaching objective
This course seeks to give an overall outline of Western architectural history, so that the students may have an in-depth understanding of the structural transition (different styles of evolution) of Western architectural history in the past 2000 years. This course hopes that students can understand the close association between the development of architectural history and the development of human civilization.
Course content
1. Generality 2. Greek Architecture 3. Roman Architecture
4. The Middle Ages Architecture
5. The Middle Ages Architectures in Italy 6. Renaissance
7. Baroque 8. American Cities 9. Nordic Romanticism
10. Catalonian Architecture 11. Avant-Garde
12. German Manufacturing Alliance and Bauhaus
13. Soviet Architecture and Cities 14. 1960's Architecture
15. 1970's Architecture 16. Answer Questions

本科三年级
外国建筑史（当代）·胡恒
课程类型：必修
学时/学分：36学时/2学分

Undergraduate Program 3rd Year
HISTORY OF WORLD ARCHITECTURE (MODERN)
• HU Heng
Type: Required Course
Study Period and Credits: 36 hours / 2 credits

教学目标
本课程力图用专题的方式对文艺复兴时期的7位代表性的建筑师与5位现当代的重要建筑师作品做一细致的讲解。本课程将重要建筑师的全部作品尽可能在课程中梳理一遍，使学生能够全面掌握重要建筑师的设计思想、理论主旨、与时代的特殊关联、在建筑史中的意义。
课程内容
1. 伯鲁乃列斯基 2. 阿尔伯蒂 3. 伯拉孟特
4. 米开朗琪罗（1） 5. 米开朗琪罗（2） 6. 罗马诺
7. 桑索维诺 8. 帕拉蒂奥（1） 9. 帕拉蒂奥（2）
10. 赖特 11. 密斯 12. 勒·柯布西耶（1）
13. 勒·柯布西耶（2） 14. 海杜克 15. 妹岛和世
16. 答疑

Teaching objective
This course seeks to make a detailed explanation to the works of 7 representative architects in the Renaissance period and 5 important modern and contemporary architects in a special way. This course will try to reorganize all works of these important architects, so that the students can fully grasp their design ideas, theoretical subjects and their particular relevance with the era and significance in the architectural history.
Course content
1. Brunelleschi 2. Alberti 3. Bramante
4. Michelangelo(1) 5. Michelangelo(2)
6. Romano 7. Sansovino 8. Palladio(1) 9. Palladio(2)
10. Wright 11. Mies 12. Le Corbusier(1) 13. Le Corbusier(2)
14. Hejduk 15. Kazuyo Sejima
16. Answer Questions

本科三年级
中国建筑史（近现代）·赵辰 冷天
课程类型：必修
学时/学分：36学时/2学分

Undergraduate Program 3rd Year
HISTORY OF CHINESE ARCHITECTURE (MODERN)
• ZHAO Chen, LENG Tian
Type: Required Course
Study Period and Credits: 36 hours / 2 credits

课程介绍
本课程作为本科建筑学专业的历史与理论课程，是中国建筑史教学中的一部分。在中国与西方的古代建筑历史课程的基础上，了解中国社会进入近代，以至于现当代的发展进程。
在对比中西方建筑文化的基础之上，建立对中国近现代建筑的整体认识。深刻理解中国传统建筑文化在近代以来与西方建筑文化的冲突与相融之下，逐步演变发展至今天成为世界建筑文化的一部分之意义。

Course description
As the history and theory course for undergraduate students of Architecture, this course is part of the teaching of History of Chinese Architecture. Based on the earlier studying of Chinese and Western history of ancient architecture, understand the evolution progress of Chinese society's entry into modern times and even contemporary age.
Based on the comparison of Chinese and Western building culture, establish the overall understanding of China's modern and contemporary buildings. Have further understanding of the significance of China's traditional building culture's gradual evolution into one part of today's world building culture under conflict and blending with Western building culture in modern times.

研究生一年级
建筑理论研究・赵辰
课程类型：必修
学时/学分：18学时/1学分

Graduate Program 1st Year
STUDIES OF ARCHITECTURAL THEORY • ZHAO Chen
Type: Required Course
Study Period and Credits: 18 hours / 1 credit

课程介绍
了解中、西方学者对中国建筑文化诠释的发展过程，理解新的建筑理论体系中对中国筑文化重新诠释的必要性，学习重新诠释中国建筑文化的建筑观念与方法。
课程内容
1. 本课的总览和基础
2. 中国建筑：西方的诠释与西方建筑观念的改变
3. 中国建筑：中国人的诠释以及中国建筑学术体系的建立
4. 木结构体系：中国建构文化的意义
5. 住宅与园林：中国人居文化意义
6. 宇宙观的和谐：中国城市文化的意义
7. 讨论

Course description
Understand the development process of Chinese and western scholars' interpretation of Chinese architectural culture, understand the necessity of reinterpretation of Chinese architectural culture in the new architectural theory system, and learn the architectural concepts and methods of reinterpretation of Chinese architectural culture.
Course content
1. Overview and foundation of this course
2. Chinese architecture: Western interpretation and the changes of western architectural concepts
3. Chinese architecture: Chinese interpretation and the establishment of Chinese architecture academic system
4. Wood structure system: The significance of Chinese construction culture.
5. Residence and garden: The cultural significance of human settlement in China
6. Harmony of cosmology: The significance of Chinese urban culture
7. Discussion

研究生一年级
建筑理论研究・王骏阳
课程类型：必修
学时/学分：18学时/1学分

Graduate Program 1st Year
STUDY OF ARCHITECTURAL THEORY • WANG Junyang
Type: Required Course
Study Period and Credits: 18 hours / 1 credit

课程介绍
本课程是西方建筑史研究生教学的一部分。主要涉及当代西方建筑界具有代表性的思想和理论，其主题包括历史主义、先锋建筑、批判理论、建构文化以及对当代城市的解读等。本课程大量运用图片资料，广泛涉及哲学、历史、艺术等领域，力求在西方文化发展的背景中呈现建筑思想和理论的相对独立性及关联性，理解建筑作为一种人类活动所具有的社会和文化意义，启发学生的理论思维和批判精神。
课程内容
第一讲：建筑理论概论
第二讲：数字化建筑与传统建筑学的分离与融合
第三讲：语言、图解、空间内容
第四讲："拼贴城市"与城市的观念
第五讲：建构与营造
第六讲：手法主义与当代建筑
第七讲：从主线历史走向多元历史之后的思考
第八讲：讨论

Course description
This course is a part of teaching Western architectural history for graduate students. It mainly deals with the representative thoughts and theories in western architectural circles, including historicism, vanguard building, critical theory, tectonic culture and interpretation of contemporary cities etc.. Using a lot of pictures involving extensive fields including philosophy, history, art, etc., this course attempts to show the relative independence and relevance of architectural thoughts and theories under the development background of western culture, understand the social and cultural significance owned by architecture as human activities, and inspire students' theoretical thinking and critical spirit.
Course content
Lect. 1: Overview of architectural theories
Lect. 2: Separation and integration between digital architecture and traditional architecture
Lect. 3: Language, diagram and spatial content
Lect. 4: "Collage city" and concept of city
Lect. 5: Tectonics and Yingzao (Ying-Tsao)
Lect. 6: Mannerism and contemporary architecture
Lect. 7: Thinking after main-line history to diverse history
Lect. 8: Discussion

研究生一年级
建筑史研究・胡恒
课程类型：选修
学时/学分：18学时/1学分

Graduate Program 1st Year
STUDIES IN ARCHITECTURAL HISTORY • HU Heng
Type: Elective Course
Study Period and Credits: 18 hours / 1 credit

教学目标
促进学生对历史研究的主题、方法、路径有初步的认识，通过具体的案例讲解使学生能够理解当代中国建筑史研究的诸多可能性。
课程内容
1. 图像与建筑史研究（1- 文学、装置、设计）
2. 图像与建筑史研究（2- 文学、装置、设计）
3. 图像与建筑史研究（3- 绘画与园林）
4. 图像与建筑史研究（4- 绘画、建筑、历史）
5. 图像与建筑史研究（5- 文学与空间转译）
6. 方法讨论 1
7. 方法讨论 2

Teaching objective
To promote students' preliminary understandings of the topic, method and approach of the historical research. To make students understand the possibilities of the contemporary study on history of Chinese architecture through specific cases.
Course content
1. Image and architectural history study (1-literature, device and design)
2. Image and architectural history study (2-literature, device and design)
3. Image and architectural history study (3- painting and garden)
4. Image and architectural history study (4- painting, architecture and history)
5. Image and architectural history study (5- literature and spatial transform)
6. Method discussion 1
7. Method discussion 2

研究生二年级
中国建构（木构）文化研究・赵辰
课程类型：必修
学时/学分：18学时/1学分

Graduate Program 2nd Year
STUDIES IN CHINESE WOODEN TECTONIC CULTURE• ZHAO Chen
Type: Required Course
Study Period and Credits: 18 hours / 1 credit

教学目标
以木为材料的建构文化是世界各文明中的基本成分，中国的木建构文化更是深厚而丰富。在全球可持续发展要求之下，木建构文化必须得到重新的认识和评价。对于中国建筑文化来说，更具有文化传统再认识和再发展的意义。
课程内容
1. 阶段一，理论基础：对全球木建构文化的重新认识
2. 阶段二，中国木建构文化的原则和方法（讲座与工作室）
3. 阶段三，中国木建构的基本形：从家具到建筑（讲座与工作室）
4. 阶段四，结构造型的发展和木建构的现代化（讲座）
5. 阶段五，建造实验的鼓动（讲座与工作室）

Teaching objective
The wood - based construction culture is the basic component of all civilizations in the world, and Chinese wood construction culture is profound and abundant. Under the requirement of global sustainable development, wood construction culture must be re-recognized and evaluated. For Chinese architectural culture, it is of great significance to re-recognize and re-develop the cultural tradition.
Course content
Stage 1, theoretical basis: re-understanding of global wood construction culture
Stage 2, principles and methods of Chinese wood construction culture (lectures and studios)
Stage 3, the basic shape of Chinese wood construction: from furniture to architecture (lectures and studios)
Stage 4, development of structural modeling and modernization of wood construction (lectures)
Stage 5, agitation of construction experiment (lectures and studios)

建筑技术课程
ARCHITECTURAL TECHNOLOGY COURSES

本科二年级
CAAD理论与实践・童滋雨
课程类型：必修
学时/学分：36学时/2学分

Undergraduate Program 2nd Year
THEORY AND PRACTICE OF CAAD • TONG Ziyu
Type: Required Course
Study Period and Credits: 36 hours / 2 credits

课程介绍

在现阶段的 CAD 教学中，强调了建筑设计在建筑学教学中的主干地位，将计算机技术定位于绘图工具，本课程就是帮助学生可以尽快并且熟练地掌握如何利用计算机工具进行建筑设计的表达。课程中整合了 CAD 知识、建筑制图知识以及建筑表现知识，将传统 CAD 教学中教会学生用计算机绘图的模式向教会学生用计算机绘制有形式感的建筑图的模式转变，强调准确性和表现力作为评价 CAD 学习的两个最重要指标。

本课程的具体学习内容包括：
1. 初步掌握 AutoCAD 软件和 SketchUP 软件的使用，能够熟练完成二维制图和三维建模的操作。
2. 掌握建筑制图的相关知识，包括建筑投影的基本概念，平立剖面、轴测、透视和阴影的制图方法和技巧。
3. 图面效果表达的技巧，包括黑白线条图和彩色图纸的表达方法和排版方法。

Course description

The core position of architectural design is emphasized in the CAD course. The computer technology is defined as drawing instrument. The course helps students learn how to make architectural presentation using computer fast and expertly. The knowledge of CAD, architectural drawing and architectural presentation are integrated into the course. The traditional mode of teaching students to draw in CAD courses will be transformed into teaching students to draw architectural drawings with sense of form. The precision and expression will be emphasized as two most important factors to estimate the teaching effect of CAD course.
Contents of the course include:
1. Use AutoCAD and SketchUP to achieve the 2-D drawing and 3-D modeling expertly.
2. Learn related knowledge of architectural drawing, including basic concepts of architectural projection, drawing methods and skills of plan, elevation, section, axonometry, perspective and shadow.
3. Skills of presentation, including the methods of expression and lay out using mono and colorful drawings

---

本科三年级
建筑技术（一）：结构、构造与施工・傅筱
课程类型：必修
学时/学分：36学时/2学分

Undergraduate Program 3rd Year
ARCHITECTURAL TECHNOLOGY 1: STRUCTURE, CONSTRUCTION AND EXECUTION • FU Xiao
Type: Required Course
Study Period and Credits:36 hours / 2 credits

课程介绍

本课程是建筑学专业本科生的专业主干课程。本课程的任务主要是以建筑师的工作性质为基础，讨论一个建筑生成过程中最基本的三大技术支撑（结构、构造、施工）的原理性知识要点，以及它们在建筑实践中的相互关系。

Course description

The course is a major course for the undergraduate students of architecture. The main purpose of this course is based on the nature of the architect's work, to discuss the principle knowledge points of the basic three technical supports in the process of generating construction (structure, construction, execution), and their mutual relations in the architectural practice.

本科三年级
建筑技术（二）：建筑物理・吴蔚
课程类型：必修
学时/学分：36学时/2学分

Undergraduate Program 3rd Year
ARCHITECTURAL TECHNOLOGY 2: BUILDING PHYSICS • WU Wei
Type: Required Course
Study Period and Credits:36 hours / 2 credits

课程介绍

本课程针对三年级学生设计，课程介绍了建筑热工学、建筑光学、建筑声学中的基本概念和基本原理，使学生能掌握建筑的热环境、声环境、光环境的基本评估方法，以及相关的国家标准。完成学业后在此方向上能阅读相关书籍，具备在数字技术方法等相关资料的帮助下，完成一定的建筑节能设计的能力。

Course description

Designed for the Grade 3 students, this course introduces the basic concepts and basic principles in architectural thermal engineering, architectural optics and architectural acoustics, so that the students can master the basic methods for the assessment of building's thermal environment, sound environment and light environment as well as the related national standards. After graduation, the students will be able to read the related books regarding these aspects, and have the ability to complete certain building energy efficiency design with the help of the related digital techniques and methods.

本科三年级
建筑技术（三）：建筑设备・吴蔚
课程类型：必修
学时/学分：36学时/2学分

Undergraduate Program 3rd Year
ARCHITECTURAL TECHNOLOGY 3: BUILDING EQUIPMENT • WU Wei
Type: Required Course
Study Period and Credits:36 hours / 2 credits

课程介绍

本课程是针对南京大学建筑与城市规划学院本科三年级学生所设计。课程介绍了建筑给水排水系统、采暖通风与空气调节系统、电气工程的基本理论、基本知识和基本技能，使学生能熟练地阅读水电、暖通工程图，熟悉水电及消防的设计、施工规范，了解燃气供应、安全用电及建筑防火、防雷的初步知识。

Course description

This course is an undergraduate class offered in the School of Architecture and Urban Planning, Nanjing University. It introduces the basic principle of the building service systems, the technique of integration amongst the building service and the building. Throughout the course, the fundamental importance to energy, ventilation, air-conditioning and comfort in buildings are highlighted.

研究生一年级
传热学与计算流体力学基础・郜志
课程类型：选修
学时/学分：18学时/1学分

Graduate Program 1st Year
FUNDAMENTALS OF HEAT TRANSFER AND COMPUTATIONAL FLUID DYNAMICS • GAO Zhi
Type: Elective Course
Study Period and Credits: 18 hours / 1 credit

课程介绍

本课程的主要任务是使建筑学/建筑技术学专业的学生掌握传热学与计算流体力学的基本概念和基础知识，通过课堂教学，使学生熟悉传热学中导热、对流和辐射的经典理论，并了解传热学和计算流体力学的实际应用和最新研究进展，为建筑能源与环境模拟打下坚实的理论基础。教学中尽量简化传热学和计算流体力学经典课本中复杂公式的推导过程，而着重于如何解决建筑能源与建筑环境中涉及流体流动和传热的实际应用问题。

Course description

This course introduces students majoring in building science and engineering / building technology to the fundamentals of heat transfer and computational fluid dynamics (CFD). Students will study classical theories of conduction, convection and radiation heat transfers, and learn advanced research developments of heat transfer and CFD. The complex mathematics and physics equations are not emphasized. It is desirable that for real-case scenarios students will have the ability to analyze flow and heat transfer phenomena in building energy and environment systems.

研究生一年级
GIS基础与应用・童滋雨
课程类型：选修
学时/学分：18学时/1学分

Graduate Program 1st Year
CONCEPT AND APPLICATION OF GIS • TONG Ziyu
Type: Elective Course
Study Period and Credits:18 hours / 1 credit

课程介绍

本课程的主要目的是让学生理解 GIS 的相关概念以及 GIS 对城市研究的意义，并能够利用 GIS 软件对城市进行分析和研究。

Course description

This course aims to enable students to understand the related concepts of GIS and the significance of GIS to urban research, and to be able to use the GIS software to carry out urban analysis and research.

研究生一年级
建筑环境学 · 郜志
课程类型：洗修
学时/学分：18学时/1学分

Graduate Program 1st Year
ARCHITECTURAL ENVIRONMENTAL SCIENCE · GAO Zhi
Type: Elective Course
Study Period and Credits:18 hours / 1 credit

课程介绍
本课程的主要任务是使建筑学／建筑技术学专业的学生掌握建筑环境的基本概念，学习建筑与城市热湿环境、风环境和空气质量的基础知识。通过课程教学，使学生熟悉城市微气候等理论，并了解人体对热湿环境的反应，掌握建筑环境学的实际应用和最新研究进展，为建筑能源和环境系统的测量与模拟打下坚实的基础。

Course description
This course introduces students majoring in building science and engineering / building technology to the fundamentals of built environment. Students will study classical theories of built / urban thermal and humid environment, wind environment and air quality. Students will also familiarize urban micro environment and human reactions to thermal and humid environment. It is desirable that students will have the ability to measure and simulate building energy and environment systems based upon the knowledge of the latest development of the study of built environment.

研究生一年级
材料与建造 · 冯金龙
课程类型：必修
学时/学分：18学时/1学分

Graduate Program 1st Year
MATERIAL AND CONSTRUCTION · FENG Jinlong
Type: Required Course
Study Period and Credits:18 hours / 1 credit

课程介绍
本课程将介绍现代建筑技术的发展过程，论述现代建筑技术及其美学观念对建筑设计的重要作用；探讨由材料、结构和构造方式所形成的建筑建造的逻辑方式；研究建筑形式产生的物质技术基础，诠释现代建筑的建构理论与研究方法。

Course description
It introduces the development process of modern architecture technology and discusses the important role played by the modern architecture technology and its aesthetic concepts in the architectural design. It explores the logical methods of construction of the architecture formed by materials, structure and construction. It studies the material and technical basis for the creation of architectural form, and interprets the construction theory and research methods for modern architectures.

研究生一年级
计算机辅助建筑设计技术 · 吉国华
课程类型：必修
学时/学分：36学时/2学分

Graduate Program 1st Year
TECHNOLOGY OF COMPUTER AIDED ARCHITECTURAL DESIGN · JI Guohua
Type: Required Course
Study Period and Credits:36 hours / 2 credits

课程介绍
随着计算机辅助建筑设计技术的快速发展，当前数字技术在建筑设计中的角色逐渐从辅助绘图转向了真正的辅助设计，并引发了设计的革命和建筑的形式创新。本课程讲授Grasshopper参数化编程建模方法以及相关的几何知识，让学生在掌握参数化编程建模技术的同时，增强以理性的过程思维方式分析和解决设计问题的能力，为数字建筑设计和数字建造打下必要的基础。
基于Rhinoceros的算法编程平台Grasshopper的参数化建模方法，讲授内容包括各类运算器的功能与使用、图形的生成与分析、数据的结构与组织、各类建模的思路与方法，以及相应的数学与计算机编程知识。

Course description
The course introduces methods of Grasshopper parametric programming and modeling and relevant geometric knowledge. The course allows students to master these methods, and enhance abilities to analyze and solve designing problems with rational thinking at the same time, building necessary foundation for digital architecture design and digital construction.
In this course, the teacher will teach parametric modeling methods based on Grasshopper, a algorithmic programming platform for Rhinoceros, including functions and application of all kinds of arithmetic units, pattern formation and analysis, structure and organization of data, various thoughts and methods of modeling, and related knowledge of mathematics and computer programming.

研究生一年级
建筑体系整合 · 吴蔚
课程类型：选修
学时/学分：18~36学时/1~2学分

Graduate Program 1st Year
BUILDING SYSYTEM INTEGRATION · WU Wei
Type: Elective Course
Study Period and Credits: 18~36 hours / 1~2 credits

课程介绍
本课程是从建筑各个体系整合的角度来解析建筑设计。首先，课程介绍了建筑体系整合的基本概念、原理及其美学观念；然后具体解读以上各个设计元素在整个建筑体系中所扮演的角色及其影响力，了解建筑各个系统之间的互相联系和作用；最后，以全球的环境问题和人类生存与发展为着眼点，引导同学们重新审视和评判我们奉为信条的设计理念和价值系统。本课程着重强调建筑设计需要了解不同学科和领域的知识，熟悉各工种之间的配合和协调。

Course description
A building is an assemblage of materials and components to obtain a shelter from external environment with a certain amount of safety so as to provide a suitable internal environment for physiological and psychological comfort in an economical manner. This course examines the role of building technology in architectural design, shows how environmental concerns have shaped the nature of buildings, and takes a holistic view to understand the integration of different building systems. It employs total building performance which is a systematic approach, to evaluate the performance of various sub-systems and to appraise the degree of integration of the sub-systems.

研究生一年级
算法设计 · 吉国华
课程类型：选修
学时/学分：18~36学时/1~2学分

Graduate Program 1st Year
ALGORITHMIC DESIGN · JI Guohuai
Type: Elective Course
Study Period and Credits: 18~36 hours / 1~2 credits

课程介绍
编程技术是数字建筑的基础，本课程主要讲授Grasshopper脚本编程和Processing编程，让学生在掌握编码编程基础技术的同时，增强以理性的过程思维方式分析和解决设计问题的能力，逐步掌握数字设计的方法，为数字设计和建造课程打好基础。

Course description
Programming technology is the foundation of digital architecture, this course mainly teaches Grasshopper script programming and Processing programming, so that students can master the basic technology of code programming, at the same time, enhance the ability to analyze and solve design problems with rational process thinking, gradually master the method of digital design, and lay a good foundation for the course of digital design and construction.

研究生二年级
建设工程项目管理·谢明瑞
课程类型：选修
学时/学分：36学时/2学分

Graduate Program 2nd Year
CONSTRUCTION PROJECT MANAGEMENT •XIE Mingrui
Type: Elective Course
Study Period and Credits: 36 hours / 2 credits

课程介绍
　　帮助学生系统掌握建设工程项目管理的基本概念、理论体系和管理方法，了解建筑规划设计在建设工程项目中的地位、特点和重要性。
　　延展建筑学专业学生基本知识结构层面，拓展学生的发展方向。

Course description
To help students systematically master the basic concept, theoretical system and management method of construction engineering project management, understand the position, characteristics and importance of architectural planning design in the construction engineering project.
To extend the basic knowledge structure level of students majoring in architecture, develop the development direction of students.

研究生三年级
建筑环境学与设计·尤伟 郜志
课程类型：必修
学时/学分：36学时/2学分

Graduate Program 3rd Year
ARCHITECTURAL ENVIRONMENTAL SCIENCE AND DESIGN• YOU Wei, GAO Zhi
Type: Required Course
Study Period and Credits:36 hours / 2 credits

课程介绍
　　本课程是基于建筑环境学课程的设计实践课程，意在将建筑环境学课程的理论知识通过设计案例的练习加以运用，加深对建筑环境学知识的理解，并训练如何通过设计优化营造良好室内环境品质。
　　课程分为授课和案例设计练习两部分，授课部分介绍目前关于被动式设计研究成果、工程实践案例中的被动式设计方法以及软件模拟分析技术。案例设计练习教授学生学习基于性能评估的优化设计方法，选取学生较为熟悉的住宅、幼儿园等体量较小的建筑类型作为设计优化对象，通过软件分析发现现有的室内环境设计不足，并基于现有研究成果知识提出优化策略，最后通过软件模拟加以验证。课程要求学生将建筑环境学课程所学知识用于本设计课程的室内环境品质的量化及控制。本课程着重训练建筑设计与环境工程学科知识的配合。

Course description
This course is a design practice course based on the course of building environment, aiming to apply the theoretical knowledge of the building environment course through the practice of design cases, so as to deepen the understanding of the knowledge of building environment, and train how to create a good indoor environment quality through design optimization.
The course is divided into two parts: teaching and case design practice. The teaching part introduces current research results of passive design, passive design methods in engineering practice cases and software simulation analysis technology.Case design practice teaches students to study optimal design methods based on performance evaluation, choose the residence, kindergarten and other buildings with small volume that students are more familiar with as the design optimization objects, find existing deficiency in indoor environment design through software analysis, propose optimization strategies according to existing research results, and finally verify through software simulation. The course requires students to apply what they have learned in the building environment course to the quantification and control of indoor environment quality in this design course.This course focuses on the integration of architectural design and environmental engineering knowledge.

研究生三年级
建筑学中的技术人文主义·窦平平
课程类型：必修
学时/学分：36学时/2学分

Graduate Program 3rd Year
TECHNOLOGY OF HUMANISM IN ARCHITECTURE • DOU Pingping
Type: Required Course
Study Period and Credits:36 hours / 2 credits

课程介绍
　　课程详尽阐释了为满足建筑的多方需求而投入的技术探索和人文关怀。课程包括四大版块，共十六个主题讲座，以案例精读的形式引介相关建筑师和学者的作品和理论。希望培养学生对建筑学中的技术议题进行批判性和人文主义的深入理解。

Course description
This course elaborates the technological endeavors and humanistic concerns in fulfilling the multifaceted architectural demands. It takes shape in a series of sixteen themed lectures, grouped in four sections, introducing prominent architects and scholars through richly illustrated case studies and interpretations. It aims to nurture the students with a critical and humanistic understanding of the role of technology playing in the discipline of architecture.

其他
MISCELLANEA

讲座
Lectures

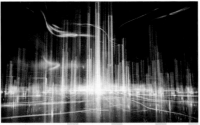

硕士学位论文列表
List of Thesis for Master Degree

研究生姓名	研究生论文标题	导师姓名
陈雪涛	环渤海地区基于寒冷气候的高密度围合院落式建筑集群防风组织设计研究	丁沃沃
杨蕾	基于建筑集群风路组织优化的珠三角高密度混合功能住区地块设计	丁沃沃
贺唯嘉	南京传统街区更新中的建筑类型学研究	丁沃沃
宋宇珺	城市肌理形态的量化方法研究——以南京为例	丁沃沃
陈安迪	结合老城区地形高差的楼宇地下空间研究——以南京大学硅巷文创中心项目设计为例	赵辰
刘怡然	与城市历史街廓更新结合的地铁出入口设计及评价研究——南京太平南路三十四标地块更新城市设计	赵辰
付伟佳	对六朝建康近郊山水人文景观的初步认知——以"周处台"项目为例	赵辰
徐瑜灵	工业建筑改造为当代青年公寓的设计方法初探	赵辰
戚迹	作为虹桥综合交通枢纽组成部分的虹桥火车站布局研究	王骏阳
李子璇	作为虹桥综合交通枢纽组成部分的虹桥国际机场发展研究	王骏阳
陆恒	基于楔形单元的可变建筑装置	吉国华
夏凡琦	测地线网壳的研究与应用初探	吉国华
曹舒琪	基于PBD的建筑群体布局生成研究——以产业园区为例	吉国华
谢军	基于四边形管状折纸的组合研究和设计初探	吉国华
张彤	基于深度学习的住宅群体排布生成实验	吉国华
章太雷	基于Kangaroo"粒子—弹簧系统"的主动弯曲结构参数化找形方法研究	吉国华
董素宏	南京审计大学莫愁校区行政楼改造设计及共享办公研究	张雷
顾妍文	西安创新港中学宿舍楼设计及公共空间研究	张雷
黄追日	云夕深澳里小白楼设计及当代乡土实践脉络研究	张雷
李江涛	南京吉山铁矿厂房改造项目——城市空间拓展与工业建筑再利用	张雷
何志鹏	南京大学大数据与人工智能科研楼设计——基于UHPC围护墙板系统的学科群建筑立面模块化设计研究	冯金龙
李汶淇	南京大学鼓楼校区图书馆改造更新设计	冯金龙
糜泽宇	南京农业大学江北校区生命科学综合大楼设计——模块化设计在学科组团建筑的应用	冯金龙
宁汇霖	南京大学军民融合研发中心设计——学科群建筑的功能组织及空间设计研究	冯金龙
王坤勇	珠三角地区基于夏季通风需求的多层居住建筑群体组合设计策略研究	周凌
杨丹	长三角地区基于风环境优化的多层板式居住建筑形体组合设计研究	周凌
张春婷	元阳哈尼族传统民居的地域建造体系	周凌
孙磊	高密度围合式住区原型的绿色性能优化研究——以沿海经济发达地区当代住宅为对象	周凌

研究生姓名	研究生论文标题	导师姓名
方 柱	我国经济发达地区大量性建筑外围护结构耐候性构造设计——接缝设计	傅 筱
孔 颖	我国经济发达地区大量性建筑耐候性构造设计——墙根与湿作业墙身设计	傅 筱
刘洋宇	我国经济发达地区大量性建筑外围护结构耐候性构造设计——洞口与干作业墙身设计	傅 筱
赵中石	我国经济发达地区大量性建筑外围护结构耐候性构造设计——墙顶与屋顶设计	傅 筱
刘信子	场所记忆视角下南京国际安全区设计研究	鲁安东
邱嘉玥	作为设计研究方法的学术策展初探——以LanD工作室的策展行动为例	鲁安东
王秋锐	基于环境建构的历史街区文化建筑设计研究——以长泾环境文化博物馆设计为例	鲁安东
陈 辰	二十世纪初新兴国家首都建设形象研究——以新德里、堪培拉、南京为例	鲁安东
何劲雁	花园洋房中国化：颐和路街区花园洋房自民国规划至今形态演变研究	鲁安东
程惊宇	日常性的影像数据库及其对建筑设计方法的启示	鲁安东
王佳倩	皮拉内西的《帕埃斯图姆的异样景观》研究——18世纪意大利建筑师视角下的"希腊复兴"	胡 恒
张培书	帕拉弟奥为波利比乌斯《历史》与凯撒《评述》所绘系列插图研究——论文艺复兴军事建筑理论的发展	胡 恒
刘晓倩	城市废弃铁路地段的活化再生研究——南京市中华门火车站场地块设计	华晓宁
万璐依	石臼湖—固城湖圩区线型村落更新研究——南京市高淳区夹埂村改造更新设计	华晓宁
顾方荣	基于天然采光优化的单层旧工业建筑办公类改造研究——以南京为例	吴 蔚
郭 超	典型城市街道街区形态的热环境分析及其对建筑能耗的影响研究	郜 志
吕 童	基于Open Modelica平台的太阳能烟囱模拟和模型控制优化研究	郜 志
马 耀	基于Modelica的家庭厨房室内污染物传播数值模拟和控制研究	郜 志
刘 晨	基于多源数据的街道空间步行适宜性评价方法优化研究——以南京主城区为例	童滋雨
杨华武	基于剖面视角面向风速的城市外部空间开放度分析——以南京为例	童滋雨
徐雅静	响应气候的独立住宅空间调节设计原理研究——以宜兴某实基地独立住宅设计为例	胡友培
秦 勤	夏热冬冷地区城市肌理形态指标与微气候的关联性规律	胡友培
张 园	从"旧式住宅"到"理想住宅"——从19世纪末至20世纪初期大众报刊与专业期刊看城市住宅变革	窦平平
潘璐梦	共生建筑——人与动物的差异化空间并置类型探索	窦平平
薛 鑫	南京近代（1840—1927年）住宅建筑立面细部特征研究及修缮策略——以汉口路22号为例	冷天、赵辰
王晓坤	闽北山区传统建造体系的当代热舒适度改进研究——董街吴宅改造烟气采暖设计	冷天、赵辰
杨颖萍	南京近代（1927—1937年）住宅建筑保护与再利用设计研究——以金银街4号为例	冷天、赵辰
王智伟	住宅室内挥发性有机化合物源散发及模拟研究	郜志、梁卫辉

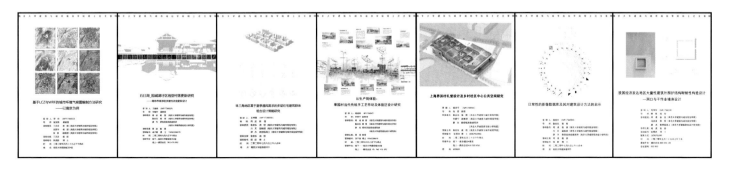

在校学生名单
List of Students

本科生 Undergraduate

2016级学生 / Students 2016

陈 帆 CHEN Fan	黄靖绮 HUANG Jingqi	丘雨辰 QIU Yuchen	余沁蔓 YU Qinman
陈靖秋 CHEN Jingqiu	黄文凯 HUANG Wenkai	石雪松 SHI Xuesong	张涵筱 ZHANG Hanxiao
陈铭行 CHEN Mingxing	雷 畅 LEI Chang	司昌尧 SI Changyao	周子琳 ZHOU Zilin
陈应楠 CHEN Yingnan	李宏健 LI Hongjian	王 路 WANG Lu	朱凌云 ZHU Lingyun
陈予婧 CHEN Yujing	李舟涵 LI Zhouhan	吴林天池 WU-Lin Tianchi	
封 翘 FENG Qiao	马彩霞 MA Caixia	吴敏婷 WU Minting	
龚之璇 GONG Zhixuan	潘 博 PAN Bo	于文爽 YU Wenshuang	

2017级学生 / Students 2017

卞直瑞 BIAN Zhirui	樊力立 FAN Lili	刘 畅 LIU Chang	沈葛梦欣 SHEN-Ge Mengxin	张凯莉 ZHANG Kaili
卜子睿 BU Zirui	甘静雯 GAN Jingwen	龙 沄 LONG Yun	沈晓燕 SHEN Xiaoyan	周金雨 ZHOU Jinyu
陈佳晨 CHEN Jiachen	顾天奕 GU Tianyi	陆柚余 LU Youyu	孙 萌 SUN Meng	周慕尧 ZHOU Muyao
陈露茜 CHEN Luxi	韩小如 HAN Xiaoru	马路遥 MA Luyao	孙 瀚 SUN Han	朱菁菁 ZHU Jingjing
陈雨涵 CHEN Yuhan	焦梦雅 JIAO Mengya	马子昂 MA Ziang	杨 帆 YANG Fan	朱雅芝 ZHU Yazhi
程科懿 CHENG Keyi	李心彤 LI Xintong	彭 洋 PENG Yang	杨佳锟 YANG Jiakun	达热亚·阿吾斯哈力 Dareya Awusihali
董一凡 DONG Yifan	林易谕 LIN Yiyu	尚紫鹏 SHANG Zipeng	杨乙彬 YANG Yibin	

2018级学生 / Students 2018

包诗贤 BAO Shixian	林济武 LIN Jiwu	邱雨欣 QIU Yuxin	肖郁伟 XIAO Yuwei	张 同 ZHANG Tong
陈锐娇 CHEN Ruijiao	刘瑞翔 LIU Ruixiang	沈 洁 SHEN Jie	熊浩宇 XIONG Haoyu	张新雨 ZHANG Xinyu
冯德庆 FENG Deqing	刘湘菲 LIU Xiangfei	宋佳艺 SONG Jiayi	徐 颖 XU Ying	周宇阳 ZHOU Yuyang
顾嵘健 GU Rongjian	陆麒竹 LU Qizhu	孙穆群 SUN Muqun	薛云龙 XUE Yunlong	阿尔申·巴特尔江 Aershen Bateerjiang
顾祥姝 GU Xiangshu	罗宇豪 LUO Yuhao	田 靖 TIAN Jing	杨 朵 YANG Duo	
何 旭 HE Xu	倪梦琪 NI Mengqi	田舒琳 TIAN Shulin	喻姝凡 YU Shufan	
李逸凡 LI Yifan	牛乐乐 NIU Lele	吴高鑫 WU Gaoxin	张百慧 ZHANG Baihui	

2019级学生 / Students 2019

高禾雨 GAO Heyu	石珂千 SHI Keqian	周昌赫 ZHOU Changhe
高赵龙 GAO Zhaolong	唐诗诗 TANG Shishi	上原舜平 SHANGYUAN Shunping
顾 靓 GU Liang	王思戎 Wang Sirong	麦吾兰江·穆合塔尔 Maiwulanjiang Muhetaer
黄辰逸 HUANG Chenyi	王智坚 WANG Zhijian	
黄小东 HUANG Xiaodong	王梓蔚 WANG Ziwei	
黄煜东 HUANG Yudong	袁 泽 YUAN Ze	
邱雨婷 QIU Yuting	张楚杭 ZHANG Chuhang	

研究生 Postgraduate

陈 硕 CHEN Shuo	董晶晶 DONG Jingjing	黄子恩 HUANG Zien	梁庆华 LIANG Qinghua	聂柏慧 NIE Baihui	王浩哲 WANG Haozhe	文 涵 WEN Han	徐新杉 XU Xinshan	臧 倩 ZANG Qian
曹永青 CAO Yongqing	高祥震 GAO Xiangzhen	季惠敏 JI Huimin	刘 刚 LIU Gang	戚 迹 QI Ji	王丽丽 WANG Lili	吴 帆 WU Fan	徐雅甜 XU Yatian	张馨元 ZHANG Xinyuan
陈思涵 CHEN Sihan	葛嘉许 GE Jiaxu	蒋玉若 JIANG Yuruo	刘江全 LIU Jiangquan	裘嘉珺 QIU Jiajun	王姝宁 WANG Shuning	吴家禾 WU Jiahe	杨瑞东 YANG Ruidong	章 程 ZHANG Cheng
陈欣冉 CHEN Xinran	耿蒙蒙 GENG Mengmeng	金璐璐 JIN Lulu	刘 宣 LIU Xuan	芮丽燕 RUI Liyan	王 坦 WANG Tan	吴 桐 WU Tong	杨 喆 YANG Zhe	赵霏霏 ZHAO Feifei
陈 妍 CHEN Yan	宫传佳 GONG Chuanjia	李鹏程 LI Pengcheng	刘姿佑 LIU Ziyou	苏 彤 SU Tong	王婷婷 WANG Tingting	吴峥嵘 WU Zhengrong	于明霞 YU Mingxia	赵媛倩 ZHAO Yuanqian
程 睿 CHENG Rui	桂 喻 GUI Yu	李恬楚 LI Tianchu	娄弯弯 LOU wanwan	孙鸿鹏 SUN Hongpeng	王一侬 WANG Yinong	谢灵晋 XIE Lingjin	于 昕 YU Xin	朱鼎祥 ZHU Dingxiang
从 彬 CONG Bin	郭嫦嫦 GUO Changchang	李 伟 LI Wei	马亚菲 MA Yafei	汤 晋 TANG Jin	王 永 WANG Yong	熊 攀 XIONG Pan	袁 一 YUAN Yi	朱凌峥 ZHU Lingzheng
代晓荣 DAI Xiaorong	黄陈瑶 HUANG Chenyao	李潇乐 LI Xiaole	梅凯强 MEI Kaiqiang	童月清 TONG Yueqing	王照宇 WANG Zhaoyu	徐 沙 XU Sha	袁子燕 YUAN Ziyan	

曹舒琪 CAO Shuqi	贺唯嘉 HE Weijia	孙 磊 SUN Lei	张培书 ZHANG Peishu	陈雪涛 CHEN Xuetao	胡慧慧 HU Huihui	刘信子 LIU Xinzi	邱嘉玥 QIU Jiayue	王 瑜 WANG Yu	杨淑婷 YANG Shuting
陈 辰 CHEN Chen	黄 煜 HUANG Yu	王佳倩 WANG Jiaqian	张 铃 ZHANG Ling	陈仲卿 CHEN Zhongqing	黄追日 HUANG Zhuiri	刘洋宇 LIU Yangyu	孙雨泉 SUN Yuquan	夏凡琦 XIA Fanqi	杨颖萍 YANG Yingping
付伟佳 FU Weijia	李子璇 LI Zixuan	干智伟 WANG Zhiwei	张 彤 ZHANG Tong	程惊宇 CHEN Jingyu	孔 颖 KONG Ying	刘怡然 LIU Yiran	谭 明 TAN Ming	徐亭亭 XU Tingting	杨泽宇 YANG Zeyu
顾方荣 GU Fangrong	刘 晨 LIU Chen	谢 军 XIE Jun	张 园 ZHANG Yuan	董素宏 DONG Suhong	李江涛 LI Jiangtao	陆 恒 LU Heng	万璐休 WAN Luyi	徐雅静 XU Yajing	张春婷 ZHANG Chunting
郭 超 GUO Chao	吕 童 LV Tong	徐瑜灵 XU Yuling	章太雷 ZHANG Tailei	方 柱 FANG Zhu	李汶祺 LI Wenqi	罗 羽 LUO Yu	王坤勇 WANG Kunyong	薛 鑫 XUE Xin	张 明 ZHANG Ming
郭 硕 GUO Shuo	马 耀 MA Yao	杨华武 YANG Huawu	赵惠惠 ZHAO Huihui	顾妍文 GU Yanwen	李雨婧 LI Yujing	糜泽宇 MI Zeyu	王秋锐 WANG Qiurui	杨 丹 YANG Dan	赵中石 ZHAO Zhongshi
韩 旭 HAN Xu	秦 勤 QIN Qin	张翱然 ZHANG Aoran	周 钰 ZHOU Yu	郭金未 GUO Jinwei	刘稷祺 LIU Jiqi	宁汇霖 NING Huilin	王 熙 WANG Xi	杨 蕾 YANG Lei	
何劲雁 HE Jinyan	宋宇珣 SONG Yuxun	张 晗 ZHANG Han	陈安迪 CHEN Andi	何志鹏 HE Zhipeng	刘晓倩 LIU Xiaoqian	潘璐梦 PAN Lumeng	王晓坤 WANG Xiaokun	杨青云 YANG Qingyun	

曹 焱 CAO Yan	陈紫葳 CHEN Ziwei	黄健佳 HUANG Jianjia	李 让 LI Rang	林瑜洋 LIN Yuyang	刘 洋 LIU Yang	时 远 SHI Yuan	王维依 WANG Weiyi	杨 璐 YANG Lu	张文轩 ZHANG Wenxuan
潮书鏞 CHAO Shuyong	迟铄雯 CHI Shuowen	黄瑞安 HUANG Ruian	李 天 LI Tian	刘婧雯 LIU Jingwen	刘颖琦 LIU Yinqi	孙晓雨 SUN Xiaoyu	王熙昀 WANG Xiyun	杨云睿 YANG Yunrui	张雅箐 ZHANG Yaxiang
陈红云 CHEN Hongyun	杜孟泽杉 DU-MENG Zeshan	蒋 健 JIANG Jian	李晓楠 LI Xiaonan	刘恺丽 LIU Kaili	罗文馨 LUO Wenxin	孙媛媛 SUN Yuanyuan	王云珂 WANG Yunke	尹子晗 YIN Zihan	张 钰 ZHANG Yu
陈健楠 CHEN Jiannan	范 勇 FAN Yong	金沛沛 JIN Peipei	李星儿 LI Xing'er	刘宛莹 LIU Wanying	倪 铮 NI Zheng	唐 萌 TANG Meng	吴慧敏 WU Huimin	张路薇 ZHANG Luwei	张 悦 ZHANG Yue
陈鹏远 CHEN Pengyuan	方园园 FANG Yuanyuan	黎飞鸣 LI Feiming	李 雅 LI Ya	刘 伟 LIU Wei	聂书琪 NIE Shuqi	王春燕 WANG Chunyan	夏心雨 XIA Xinyu	张珊珊 ZHANG Shanshan	郑 航 ZHENG Hang
陈启宁 CHEN Qining	冯时雨 FENG Shiyu	李谷羽 LI Guyu	梁晓蕊 LIANG Xiaorui	刘 霄 LIU Xiao	秦 岭 QIN Ling	王慧文 WANG Huiwen	徐琳茜 XU Linxi	张思琪 ZHANG Siqi	周珏伦 ZHOU Juelun
陈 晓 CHEN Xiao	郭 鑫 GUO Xin	李家祥 LI Jiaxiang	林晨晨 LIN Chenchen	刘晓芬 LIU Xiaofen	邵夏梦 SHAO Xiameng	王少君 WANG Shaojun	徐志超 XU Zhichao	张 涛 HANG Tao	周 郅 ZHOU Zhi
陈妍霓 CHEN Yanni	胡名卉 HU Minghui	李斓珺 LI Lanjun	林 宇 LIN Yu	柳 妍 LIU Yan	施少銎 SHI Shaojun	王 顺 WANG Shun	严华东 YAN Huadong	张文欣 ZHANG Wenxin	左 斌 ZUO Bin

卞 真 BIAN Zhen	程 绪 CHENG Xu	戈可辰 GE Kechen	黄晓寻 HUANG Xiaoxun	李嘉伟 LI Jiawei	刘 贺 LIU He	宋晓宇 SONG Xiaoyu	王鹏程 WANG Pengcheng	辛 宇 XIN Yu	周诗琪 ZHOU Shiqi
岑国桢 CEN Guozhen	程 慧 CHENG Yi	谷雨阳 GU Yuyang	皇甫子玥 HUANGFU Ziyue	李心仪 LI Xinyi	罗逍遥 LUO Xiaoyao	谭路路 TAN Lulu	王若辰 Wang Ruochen	杨子媛 YANG Ziyuan	朱 硕 ZHU Shuo
车娟娟 CHE Juanjuan	戴添趣 DAI Tianqu	顾梦婕 GU Mengjie	孔 严 KONG Yan	李 乐 LI Yue	吕文倩 LV Wenqian	孙 其 SUN Qi	王赛施 WANG Saishi	袁 琴 YUAN Qin	朱维韬 ZHU Weitao
陈莉莉 CHEN Lili	丁珺宸 DING Zhantu	何 璇 HE Xuan	匡 鑫 KUANG Xin	李 雪 LI Xue	明文静 MING Wenjing	谭锦楠 TAN Jinnan	王 旭 WANG Xu	臧 哲 ZANG Zhe	朱晓晨 ZHU Xiaochen
陈星雨 CHEN Xingyu	董 青 DONG Qing	胡应航 HU Yinghang	况 赫 KUANG He	李伊萌 LI Yimeng	莫 默 MO Mo	王新强 WANG Xinqiang	王子涵 WANG Zihan	张 含 ZHANG Han	
陈一帆 CHEN Yifan	范嫣琳 FAN Yanlin	洪 静 HONG Jing	李 昂 LI Ang	李芸梦 LI Yunmeng	濮文睿 PU Wenrui	王家洲 WANG Jiazhou	魏雪仪 WEI Xueyi	张 尊 ZHANG Zun	
陈玉珊 CHEN Yushan	冯杨帆 FENG Yangfan	侯自忠 HOU Zizhong	李博雅 LI Boya	梁 颖 LIANG Ying	沈育辉 SHEN Yuhui	王 锴 WANG Kai	温 琳 WEN Lin	赵 彤 ZHAO Tong	
陈志凡 CHEN Zhifan	傅婷婷 FU Tingting	黄 菲 HUANG Fei	李昌曦 LI Changxi	廖伟平 LIAO Weiping	史鑫尧 SHI Xinyao	王梦兰 WANG Menglan	翁 昕 WENG Xin	郑经纬 ZHENG Jingwei	

图书在版编目（CIP）数据

南京大学建筑与城市规划学院建筑系教学年鉴.2019—2020/胡友培编. -- 南京：东南大学出版社，2020.12
ISBN 978-7-5641-9317-1

Ⅰ.①南… Ⅱ.①胡… Ⅲ.①南京大学—建筑学—教学研究—2019—2020—年鉴 Ⅳ.①TU-42

中国版本图书馆CIP数据核字（2020）第257989号

南京大学建筑与城市规划学院建筑系教学年鉴 2019—2020
Nanjing Daxue Jianzhu Yu Chengshi Guihua Xueyuan Jianzhuxi Jiaoxue Nianjian 2019—2020

编　　者：胡友培
编 委 会：丁沃沃　赵　辰　吉国华　周　凌　胡友培
版面制作：胡友培　孙　林　王峥涛　燕海南　李　让　李　雪
封面设计：李　雪
责任印制：周荣虎
责任编辑：姜　来　魏晓平

出版发行：东南大学出版社
社　　址：南京市四牌楼2号
出 版 人：江建中
网　　址：http://www.seupress.com
邮　　箱：press@seupress.com
邮　　编：210096
经　　销：全国各地新华书店
印　　刷：南京新世纪联盟印务有限公司
开　　本：889 mm×1194 mm　1/20
印　　张：8.5
字　　数：352千
版　　次：2020年12月第1版
印　　次：2020年12月第1次印刷
书　　号：ISBN 978-7-5641-9317-1
定　　价：68.00元

本社图书若有印装质量问题，请直接与营销部联系。电话：025-83791830

ISBN 978-7-5641-9317-1

定价：68.00元

NJU SA 2014-2015

THE YEAR BOOK OF ARCHITECTURE PROGRAM SCHOOL OF ARCHITECTURE AND URBAN PLANNING

南京大学建筑与城市规划学院建筑系　教学年鉴

王 丹 丹 编　　EDITOR: WANG DANDAN

东南大学出版社·南京 ∕ SOUTHEAST UNIVERSITY PRESS, NANJING